Main Command Post-Operational Detachments (MCP-ODs) and Division Headquarters Readiness

Stephen Dalzell, Christopher M. Schnaubelt, Michael E. Linick, Timothy R. Gulden, Lisa Pelled Colabella, Susan G. Straus, James Sladden, Rebecca Jensen, Matthew Olson, Amy Grace Donohue, Jaime L. Hastings, Hilary A. Reininger, Penelope Speed

Prepared for the United States Army

For more information on this publication, visit www.rand.org/t/RR2615

Library of Congress Cataloging-in-Publication Data is available for this publication.
ISBN: 978-1-9774-0225-7

Published by the RAND Corporation, Santa Monica, Calif.

Preface

This report documents research and analysis conducted as part of a project entitled Multi-Component Units and Division Headquarters Readiness sponsored by U.S. Army Forces Command. The purpose of the project was to identify the effects of implementing the multicomponent unit (MCU) program for all division headquarters (HQ) on their readiness to rapidly respond to contingencies and their ability to conduct mission command from alert through completed deployment in theater. The project also sought to develop potential mitigation strategies, as needed, for division HQs based in the continental United States (CONUS).

The Project Unique Identification Code (PUIC) for the project that produced this document is HQD177474.

This research was conducted within RAND Arroyo Center's Personnel, Training, and Health Program. RAND Arroyo Center, part of the RAND Corporation, is a federally funded research and development center (FFRDC) sponsored by the U.S. Army.

RAND operates under a Federal-Wide Assurance (FWA00003425) and complies with the Code of Federal Regulations for the Protection of Human Subjects Under United States Law (45 CFR 46), also known as "the Common Rule," as well as with the implementation guidance set forth in Department of Defense (DoD) Instruction 3216.02. As applicable, this compliance includes reviews and approvals by RAND's Institutional Review Board (the Human Subjects Protection Committee) and by the U.S. Army. The views of sources utilized in this study are solely their own and do not represent the official policy or position of DoD or the U.S. government.

Contents

Figures

Tables

Summary

In July 2013, then–Secretary of Defense Charles Hagel directed a 20-percent reduction in spending on management-level headquarters (HQ).[1] The Secretary of the Army and the Chief of Staff of the Army (CSA) subsequently directed the creation of Focus Area Review Groups (FARGs) to "develop bold executable recommendations" and explore 25-percent reductions in institutional and operational HQ. The unit structures that eventually resulted from this direction became known as the "FARG II" HQ design.[2] This design included the creation of a new unit type called the Main Command Post–Operational Detachment (MCP-OD), through which reserve component (RC) personnel would augment active component (AC) division and corps staffs. Figure S.1 shows the previous, all-AC structure and the two subsequent modifications.

U.S. Army Forces Command (FORSCOM) asked RAND Arroyo Center to identify the effects of this design on division HQ readiness to respond rapidly to contingencies and their

Figure S.1
Focus Area Review Group II Reductions to Division Headquarters

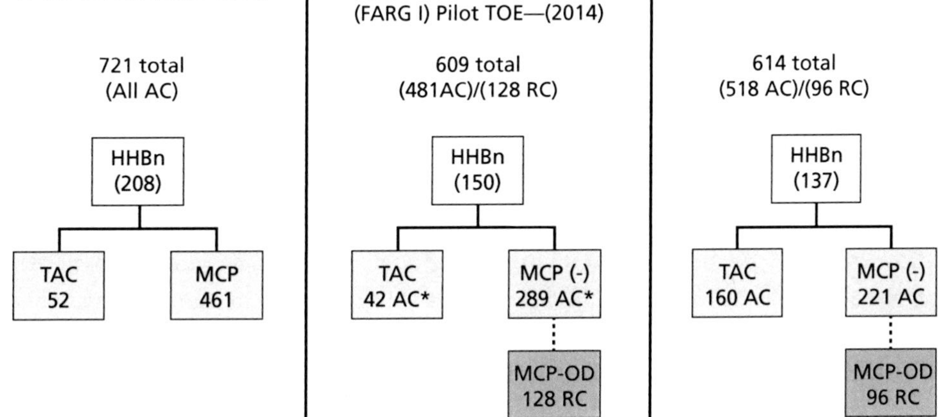

SOURCE: U.S. Army Combined Arms Center, "Corps and Division Redesign (FARG II) Force Design Update Brief," December 2015.
KEY: HHBn = headquarters and headquarters battalion; TAC = tactical command post; TOE = table of organization and equipment.

---

1 Ashton B. Carter, "20% Headquarters Reductions," USNI.org, August 2, 2013.

2 U.S. Army Combined Arms Center, 2015, Slide 4.

ability to conduct mission command from alert through completed deployment in theater. FORSCOM further asked RAND Arroyo Center to develop potential mitigation strategies, as appropriate.

Methodology

This project first reviewed U.S. Army history, doctrine, and force structure documents to define the conditions under which a division HQ can be expected to deploy and the missions it might be assigned. As one might expect, this research showed that doctrine describes at least four missions, all of which have been executed by at least one division HQ since Operation Desert Storm and the end of the Cold War. The review of the FARG II design decision highlighted the fact that many of the concerns currently being expressed about the division design were acknowledged as areas of risk during the decisionmaking process.

Next, the project team visited several AC divisions during either premobilization warfighter exercises (WFX) or following a deployment to interview participants about their experiences in such a multicomponent HQ. We supplemented these visits with telephone interviews with other units, particularly with Army National Guard (ARNG) units still "standing up" their MCP-ODs. The team then used a DOTMLPF-P (doctrine, organization, training, materiel, leadership, personnel, facilities, and policy) framework to analyze how AC divisions and RC commands have attempted to prepare their combined forces for employment as a single HQ and where their efforts have been challenged.

The team then developed a model to illustrate the conditions under which a division HQ will have increased risk as it tries to meet combatant commander requirements on a desired time line. The model enabled us to vary key parameters of a deployment scenario, including the length of notification time before mobilization, number and skills of HQ personnel required for the mission, and the assigned strength percentages and duty military occupational specialty qualification (DMOSQ) rates.

Major Findings

In the limited number of deployments executed thus far, MCP-ODs have demonstrated that, if they are given at least 270 days of advance notice, they can successfully deploy with an AC division HQ and accomplish their missions. However, without substantial advanced notification of sourcing, MCP-OD personnel will not be able to deploy as quickly as the AC soldiers in a division's command posts. This limitation was known when FARG II was designed and explicitly accepted as a risk by the CSA. Stakeholders may not like the existence of this risk, but, given the imperative to reduce the size of the Army, the cuts to division HQ structures were accepted by the Army's leadership as a trade-off to preserve force structure elsewhere.

Although we fleshed out additional details on the risks, such as the probable time lines for MCP-OD availability, the instances of risks found in our research—to include discussions with numerous stakeholders and multiple echelons—were adequately anticipated by the FARG II designers: the types of mitigation of the FARG I risks that the FARG II design and the MCP-OD design were supposed to address do in fact deal with them. Nonetheless, room for improvement exists in both design and execution. We found several implications of MCP-OD readiness and availability limitations that might be mitigated.

After assuming various levels of MCP-OD readiness, mission requirements, and deployment time lines, our model showed that the new structure should be able to meet a range of rotational and small-scale contingencies. However, in the worst-case scenario of a full HQ deployment on less than 90 days' notice, a main command post (MCP) shortfall of approximately 24 RC personnel emerged. This gap might be mitigated through a short-term "bridge" of 24 AC personnel from outside the HQ, two-thirds of whom would be backfilled by RC soldiers and returned to their parent units before the HQ deployed.

Recommendations

On one hand, the above results could reassure Army planners, because the HQ shortfalls under the most likely scenarios seem manageable. On the other hand, we recognize that Army planners do not live in a world of "good enough," and there are reasons why a specific MCP-OD deployment might not go as well as our model predicts. For example, one state might have more trouble than others in recruiting, training, and retaining some of the military occupational specialties (MOSs) in its MCP-OD, or the division may have recently returned from another deployment, creating temporary shortfalls in the number of soldiers assigned and deployable. Planners, from FORSCOM down to the divisions, would be expected to plan for the worst case and ask what can be done to minimize its effects.

For these reasons, we recommend the Army consider a range of actions that we expect will improve the likelihood of success and minimize the risk of such HQ shortfalls when called to deploy. These recommendations include the following:

Division Structure and Manning

1. The Army may want to consider two different division designs: one fully manned by the AC, focused on short-notice deployments across the spectrum of conflict and without a MCP-OD, and one that accepts the risks of the FARG II design (as mitigated by the MCP-OD).
2. The Army should also reconsider creating division HQ as true multicomponent units (MCUs) and integrating the MCP-ODs accordingly, versus the current designation as partner units.
3. Training and Doctrine Command (TRADOC) should document MCP-OD civil affairs (CA) and military information support operations (MISO) requirements on one U.S. Army Reserve (USAR) unit table of organization and equipment (TOE) and modified table of organization and equipment (MTOE). Pending a more comprehensive redesign of CA and MISO TOEs, the easiest near-term solution might be to follow current practice and document those positions in a distinct cell within U.S. Army Civil Affairs and Psychological Operations Command (USACAPOC).
4. FORSCOM and the ARNG should examine split stationing options for MCP-ODs that will better align soldiers for both career progression and for appropriate training opportunities.

5. If, after several years, the MCP-ODs have consistent difficulties filling some positions due to the lack of sustainable personnel pipelines in their states, the MCP-OD MTOE could be converted to a true multicomponent unit and USAR soldiers assigned to fill the positions based on that component's core competencies.
6. Each division should develop a command post contingency staffing plan for filling critical position shortfalls if they are not sufficiently mitigated by their partnered MCP-ODs.

Doctrine and Guidance

7. FORSCOM should consider promulgating an information paper or other communication on Army decisionmaking and division HQ force structure trade-offs. TRADOC should consider including division HQ design in the curriculum for intermediate-level professional military education courses. The division HQ TOE (and MTOE) narratives should more explicitly lay out the FARG II design risk and the relationship between the division HQ and the MCP-OD.
8. FORSCOM should consider forming a study team to consider how Objective T readiness reporting might be better applied to MCP-OD Commander's Unit Status Report (CUSR) reporting than the current process.
9. The Combined Arms Doctrine Directorate, U.S. Army Combined Arms Center should incorporate MCP-OD considerations in the forthcoming revision of ATP 6-0.5, *Command Post Organization and Operations.*
10. FORSCOM should consider forming a study group to assess garrison support requirements for division HQ in the wake of FARG II reductions. For example, can additional tasks be contracted or assigned to echelons below division? Should some of these requirements be the responsibility of Installation Management Command versus being tasked to divisions?

MCP-OD Training and Resourcing

11. The Army should consider designating one or more MCP-ODs as Focused Readiness Units and resourcing them to enable deployment within 60 days of notification.
12. AC divisions with MCP-ODs should collaborate closely with MCP-OD commanders, the appropriate state Joint Forces Headquarters, and the U.S. Army Reserve Command as appropriate to (a) plan, execute, and validate training as required to meet deployment requirements within 270 days of notification of sourcing, (b) identify and document shortfalls given current RC resourcing levels, and (c) plan for postmobilization training requirements to close the identified shortfalls.
13. Division Chiefs of Staff and MCP-OD commanders should collaborate closely to synchronize AC and RC training management cycles to optimize MCP-OD readiness and integration into the division HQ. One approach to maximize cross-component training would be to have some AC soldiers in selected division command post sections periodically train on weekends to coincide with MCP-OD inactive duty training and be given a four-day training holiday in exchange.

Acknowledgments

We thank our primary action officers at FORSCOM, first CW5 John A. Robinson and then Kristin Blake, for their support, advice, and assistance with this project. Their help was critical in our obtaining key data and identifying the right people to interview.

This project benefitted from the expertise, experience, and energy of many of our RAND colleagues, including Pardee RAND Graduate School Fellows Elizabeth Bartels and Ben Smith, Army Fellows Rob Federigan and Chuck Douglas, Abigail Casey, Daniel Ibarra, Josh Klimas, and Laurie McDonald. We owe a particular debt of gratitude to the administrative assistants to the primary investigators: Mark Hvizda, Rhonda Normandin, and Nina Ryan.

We particularly appreciate the comments and advice from our reviewers, Michael Meese of American Armed Forces Mutual Aid Association and Raphael Cohen of RAND. Their input resulted in sharper analysis and more cogent explanations of our findings.

The project could not have succeeded without the cooperation of soldiers and civilians working for all three Army components. In our visits to five Army installations, our points of contact were consistently helpful and went out of their way to ensure we could meet with everyone from division commanders to the most junior soldiers.

Abbreviations

101st Airborne	101st Airborne Division (Air Assault)
10th Mountain	10th Infantry Division (Mountain)
1st ID	1st Infantry Division
25th ID	25th Infantry Division
3rd ID	3rd Infantry Division
82nd Airborne	82nd Airborne Division
AAR	after-action review
AC	active component
ACE	all-source collection element
ADRP	Army Doctrine Reference Publication
AGR	Active Guard/Reserves
AMD	Air and Missile Defense
AOC	Army Operating Concept
ARFOR	Army Forces
ARNG	Army National Guard
AT	annual training
ATP	Army Techniques Publication
BCT	Brigade Combat Team
CA	Civil Affairs
CAC	Combined Arms Center
CBRN	chemical, biological, radiological, nuclear
CD	Cavalry Division
CDR	commander

CJCS	Chairman of the Joint Chiefs of Staff
CMD	command
COE	center of excellence
CP	Command Post
CPOF	Command Post of the Future
CSA	Chief of Staff of the Army
CTG	command training guidance
CUOPS	current operations
CUSR	Commander's Unit Status Report
CV	coefficient of variation
DCGS-A	Distributed Common Ground System–Army
DIV	division
DMOSQ	duty military occupational specialty qualification
DoD	Department of Defense
DOTMLPF-P	doctrine, organization, training, materiel, leadership, personnel, facilities, and policy
DREAR	division rear command post
DTAC	division tactical command post
ELM	element
FARG	Focus Area Review Group
FM	field manual
FORSCOM	U.S. Army Forces Command
FTNGD	full-time National Guard duty
FUOPS	future operations
FY	fiscal year
GRF	Global Response Force
HADR	humanitarian assistance and disaster relief
HCLOS	High-capacity line of sight
HHBn	headquarters and headquarters battalion
HHC	headquarters and headquarters company
HQ	headquarters

HQDA	Headquarters, Department of the Army
HR	human resources
ID	inactive duty
IDT	inactive duty training
INSTL-MNT	installer-maintainer
IT	information technology
JFC	joint force command
JTF	joint task force
MCCOE	Mission Command Center of Excellence
MCP	main command post
MCP-OD	Main Command Post–Operational Detachment
MCTP	Mission Command Training Program
MCU	multicomponent unit
MEB	Maneuver Enhancement Brigade
METL	mission-essential task list
MI	military intelligence
MISO	military information support operations
MOS	military occupational specialty
MTOE	modified table of organization and equipment
MVR	maneuver
NCO	noncommissioned officer
NTL	no later than
OPS	operations
PAO	public affairs office
PR	personnel recovery
RA	Regular Army
RC	reserve component
SACP	Support Area Command Post
SECDEF	Secretary of Defense
SCTY COOP	security cooperation

SIGINT	signals intelligence
SJA	staff judge advocate
SOP	standard operating procedure
TAC	tactical command post
TAPDB	Total Army Personnel Database
TDA	table of distribution and allowances
TOE	table of organization and equipment
TRADOC	Training and Doctrine Command
USAR	U.S. Army Reserve
UTP	unit training plan
WFX	warfighter exercise

CHAPTER ONE

Introduction

Who Asked Us to Do What and Why It Is Important

In July 2013, then-Secretary of Defense Charles Hagel directed a 20-percent reduction in spending on management-level headquarters (HQ).[1] The Secretary of the Army and the Chief of Staff of the Army (CSA) subsequently directed the creation of Focus Area Review Groups (FARGs) to "develop bold executable recommendations" and explore 25-percent reductions in institutional and operational HQ. In late January 2015, the Maneuver Center of Excellence completed a design for single component corps and division HQ at strengths of 570 and 520 soldiers, respectively, shown in Figure 1.1 below.[2]

This initial design (FARG I) accepted a known risk to HQ capacity in case of a contingency or other short-notice deployment.[3] The CSA issued guidance on February 3, 2015, to develop a multicomponent design to mitigate some of the risks identified in the single-component designs. The unit structures resulting from this direction became known as "FARG II."[4] While the specific structure of division HQ is rarely fixed for long, and future Army budgets and operational requirements may present opportunities to reverse some of the FARG II changes—or even because of these potential changes—it is important to understand the innovative features of the FARG II model and assess the actual level of risk the Army is experiencing in applying it.

The FARG II design included a new unit type called the Main Command Post–Operational Detachment (MCP-OD). These units were created to provide reserve component (RC) personnel to augment active component (AC) division and corps staffs, in the form of individuals who will fill specific positions throughout the staff sections of the command post and the HQ battalion when they are mobilized. The MCP-OD itself has the structural attributes of a unit, including a separate modified table of organization and equipment (MTOE) and requirement for readiness reporting, but it only functions as such in peacetime. Once it merges with the AC portion of the main command post (MCP), the larger HQ assumes its command and support functions.

---

[1] Ashton B. Carter, "20% Headquarters Reductions," USNI.org, August 2, 2013.

[2] U.S. Army Combined Arms Center, "Corps and Division Redesign (FARG II) Force Design Update Brief," December 2015, Slides 3–4. The process that led to 25-percent reductions in the size of headquarters and creation of the MCP-ODs is detailed in Chapter Two.

[3] Interviews with doctrine writers and directors, deputy directors, and branch chiefs at the Combined Arms Center who had been involved with FARG design and implementation, January 4–5, 2017, Ft. Leavenworth, Kan.

[4] U.S. Army Combined Arms Center, 2015, Slide 4.

Figure 1.1
Focus Area Review Group II Reductions to Division Headquarters

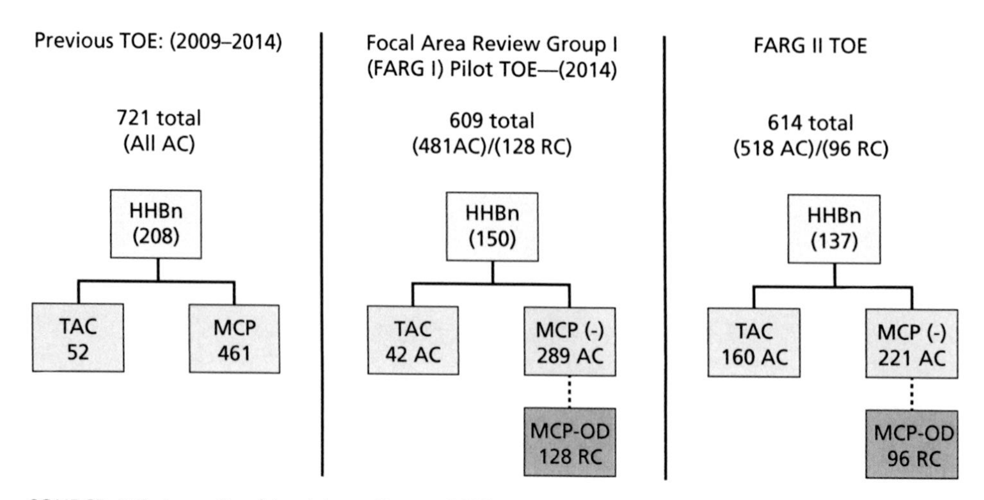

SOURCE: U.S. Army Combined Arms Center, 2015.
KEY: HHBn = headquarters and headquarters battalion; TAC = tactical command post; TOE = table of organization and equipment.

For several reasons, the MCP-OD still has a degree of operational risk.[5] In addition to the reduction in total authorizations for the HQ (compared with the pre-FARG structure), RC personnel may not be as readily available as AC soldiers assigned to a command post, would probably be available for fewer days per deployment because of policy restrictions on the amount of time a reserve soldier can be mobilized, and would be eligible for fewer deployments during the period of assignment to a MCP-OD. At the same time, because a division HQ does not always deploy with all personnel authorized on its MTOE or repeatedly for the same mission, it is unclear how often and to what extent the lesser availability of the RC personnel will affect HQ operations.[6]

In light of these issues, U.S. Army Forces Command (FORSCOM) asked RAND Arroyo Center to identify the effects of partnering each AC division HQ stationed in the United States with an MCP-OD in terms of their readiness to respond rapidly to contingencies and their ability to conduct mission command from alert through completed deployment in theater. FORSCOM further asked RAND Arroyo Center to develop potential mitigation strategies, as appropriate.

Key Questions

As we began our research, we learned that the developers of the FARG II design had explicitly recognized numerous risks and that these had been accepted by the CSA. We therefore focused our research on the following questions:

---

[5] The focus of this study is the impact upon AC division headquarters, but each of the Army National Guard (ARNG) divisions were also slated to be converted to the new structure and assigned an MCP-OD.

[6] See Secretary of Defense, Memorandum for Secretaries of the Military Departments, Chairman of the Joint Chiefs of Staff, and Under Secretaries of Defense, "Utilization of the Total Force," January 19, 2007.

- Are there significant risks with FARG II that were *not* identified during the design process and thus unknowingly accepted by Army leadership?
- What additional steps might be taken to further mitigate both anticipated and unanticipated risks?

Assumptions

At the outset, this study began with a series of assumptions. The ones that remain noteworthy for this report include the following:

- The Army was not asking for a review of the new division HQ or the MCP-OD concept per se. In several rounds of discussions, FORSCOM confirmed their question was how to work within the structure that had been developed. RAND Arroyo Center could recommend slight changes in the grade/duty military occupational specialty/section structure of the division HQ and the MCP-OD if justified by the data.
- There is no *essential* difference in the quality of soldiers in the three components. Whatever differences exist in the job performance of individuals reflect differences in their education, training, and the amount of time they have served in their current position or ones like it. Unless explicitly stated, we assumed an interview subject expressing a preference for AC soldiers was referring to their constant availability and geographic proximity, not their personal attributes.
- There is no single division mission (e.g., "Deploy the full division HQ by airlift within X days to command three Brigade Combat Teams [BCTs] in major ground operations in desert and urban terrain against a near-peer competitor within Y days") against which HQ readiness can be assessed. Therefore, readiness needed to be weighed against a variety of missions.

Desired Outcome of Study

The major challenge addressed in this study concerns the consequences brought about by the Army's decision to cut division HQ authorizations and then to partially replace those cuts with RC soldiers. Interestingly, these MTOE changes were made to all division MTOEs, including ARNG divisions, in a "one size fits all" solution. Consistent with the scope of this study, we did not examine the ARNG divisions, having assumed that in such cases all the soldiers (MCP-OD and the rest of the division HQ) have the same readiness levels. We therefore analyzed the impact only on AC divisions.

We were unable to develop a clear distinction in our interviews, but it seemed that in some cases interviewees did not understand the sequence of cuts and backfills with RC soldiers. Some may have thought the real question was concerning AC soldiers compared to RC soldiers, rather than AC compared to nobody in those slots. But, as will be seen, in most cases AC leaders thought RC fills were better than no soldiers at all. In only one case did an AC leader say that having nobody would be better than having an RC MCP-OD, although some AC staff officers and noncommissioned officers (NCOs) made critical comments.[7]

---

[7] AC field-grade officer, June 14, 2017, Ft. Bragg, N.C.; AC staff officer, February 15, 2017, Ft. Bliss, Tex.; AC NCO, June 14, 2017, Ft. Bragg, N.C.

We took it as a given that, ceteris paribus, every AC division would prefer being all AC soldiers, with enough of them to meet all contingencies, but that this was not an option, at least within the scope of this study. Instead, we focus upon the readiness issues of the current HQ under this sourcing solution and how they might best be mitigated.

Methodology

To answer our key questions, the RAND Arroyo Center research team followed multiple lines of inquiry: a literature review, an analysis of empirical data, interviews with stakeholders, and modeling the readiness of the consolidated HQ under various scenarios.

The first step toward analyzing the FARG II changes and the MCP-ODs was to understand the division HQ in the U.S. Army. This task includes reviewing the history of the division as a combat formation, how its organization and roles have changed over time, and the effects of those changes on the HQ element. This task also includes understanding how the divisions and their HQ were actually used in combat and noncombat deployments, not just how they were designed and expected to be used. Thus, we began with a literature review of division and command post doctrine and a data call on such topics as division histories, MCP-OD deployments, and personnel fill and duty military occupational specialty qualification (DMOSQ) rates in MCP-ODs and AC division HQ.

At the same time, we interviewed approximately 90 subject-matter experts at FORSCOM, the U.S. Army Combined Arms Center (CAC), and selected division HQ and MCP-ODs. Approximately one-sixth of the interview subjects were from the RCs, mostly the ARNG. In coordination with the sponsor, we chose warfighter exercises (WFX) as the primary venue for interviews with soldiers in division HQ and MCP-ODs. This seemed to provide a cost-effective way to meet with both soldiers assigned to MCP-ODs and those AC soldiers who work with MCP-ODs. This focus on WFX turned out to be problematic for the following reasons:

- In one case, the MCP-OD had not been activated yet.
- In other cases, the MCP-ODs were newly formed and had not yet reached initial operating capability.
- In virtually every case we observed, the division HQ was also augmented from subordinate AC units and other sources and thus may have not felt the full effect of the conversion of MCP-OD positions.
- In the case of the 101st Airborne Division (Air Assault),[8] our interviews occurred shortly after the deployment, not before it. It had used a onetime multicomponent unit model, so some of its experience may not apply to the FARG II or to MCP-ODs as enduring units.[9]
- Another arguable outlier was the 25th Infantry Division (25th ID), which (1) had not yet converted to the FARG II MTOE at the time of their WFX but evinced having given

[8] Hereafter referred to as 101st Airborne.

[9] Nonetheless, a key aspect of Army culture is that a "real-world deployment" carries more weight than an exercise like WFX, and we therefore included the 101st Airborne in our sample.

considerable thought about it and (2) will have a U.S. Army Reserve (USAR) MCP-OD rather than one from the ARNG.[10]
- The WFX we observed constituted a convenience sample rather than a random sample. These were simply the set of events that took place during the roughly nine-month data collection phase of this project. Interviews were based upon the availability of MCP-OD leaders, and officers and senior NCOs in the division command posts. Because neither the WFX nor the interviews conducted concurrently were a random sample, we cannot statistically estimate the representativeness of this sample compared to the population of MCP-ODs.

Nonetheless, we believe the WFX were an effective opportunity to interview a significant number of soldiers directly concerned with whether the MCP-OD enterprise is a success, at the point where they were wrestling with the challenges standing in its way. These interviews provided numerous observations that contributed to our analysis. We also conducted several interviews outside of the WFX context through phone and office visits.

To complement the interviews and empirical data, we developed a model of MCP-OD readiness under varying lengths of notification time before mobilization, number and skills of HQ personnel required for the mission, and assigned strength percentages and DMOSQ rates. Because the most recent personnel data set available to us ended on September 30, 2016, we used this model to estimate readiness status during the later period of this research as well as readiness under different scenarios.

Road Map to the Report

Chapter Two provides a brief history of division deployments and changes to their structure over time and describes the changes to the number of personnel authorized in a division HQ and the resulting establishment of MCP-ODs. This analysis provides context for the subsequent assessment of the effect of FARG II changes across a wide range of potential division HQ missions.

In Chapter Three, we analyze the current state of the division HQ and the establishment of MCP-ODs. Using the DOTMLPF-P (doctrine, organization, training, materiel, leadership, personnel, facilities, and policy) framework, we present the results of interviews, observation, and other research on the readiness of the current HQ to deploy and accomplish their missions, focusing on those elements of the HQ that rely on the contributions of the MCP-ODs.

The results of modeling and the implications of short-notice missions for MCP-OD and division HQ readiness are presented in Chapter Four. There, the report focuses on presenting a quantitative model that allowed the team to show how shortfalls in manning and training interact with requirements (varying in terms of both numbers and skills required and the time line for their usage). Thus, we illustrate how differing assumptions about the availability of

---

[10] Having a USAR MCP-OD probably offers advantages (the unit can be manned by soldiers from different states and does not have to worry about competing requirements from its state chain of command) and disadvantages (because the Army Reserve has few combat arms units, it may have more challenges filling combat arms military occupational specialties [MOSs] in the MCP-OD). The key point is simply that the challenges facing the MCP-OD for the 25th ID will differ from those facing ARNG ones. Further, we note that corps of MCP-ODs are drawn from the USAR but are specifically beyond the scope of this study.

trained and ready personnel and the conditions for their deployment could lead to dramatically different findings about whether a MCP-OD would be up to the task.

The concluding chapter summarizes the findings of the research and offers recommendations for actions the Army could take to reduce further the operational risk brought about by the FARG II design and the creation of the MCP-ODs.

Notes on Terminology

One challenge in writing about this topic is that the Army has very specific terminology for different relationships and structures. This report began life as a study on the "multicomponent" division HQ, but, strictly speaking, a multicomponent unit (MCU) would have a single unit identification code (UIC) but be filled by individuals from more than one component. Because the current division HQ is actually two entities, a division HQ (AC or ARNG) and an MCP-OD (ARNG or USAR), it is not an MCU. To avoid confusion, in this report we will use "consolidated HQ" when we are referring to the full division HQ as it appears when the two UICs combine to function as a multicomponent entity.[11]

In accordance with current usage, "reserve component" (RC) is used when referring to both the ARNG and the USAR and applies to both the unit structure and the personnel in each. "Active component" (AC) refers to that part of the Army *structure*, but "regular Army" (RA) is used for the category of personnel that generally serve in these AC units in this component.

---

[11] The differences between multicomponent units, associations, and partnerships are defined in Headquarters, United States Army Forces Command, FORSCOM Regulation 220-2, *Methods for Integrating Regular Army, Army National Guard, and Army Reserve Organizations*, Ft. Bragg, N.C., May 31, 2017. It states: "Main Command Post-Operational Detachment (MCP-OD) is a special example of the partnership method. The SECARMY [Secretary of the Army] delayed associating MCP-ODs until the Associated Units Pilot (AUP) is completed in fiscal year (FY) 2019. While currently indicated in the Force Management System Website (FMSWeb) where MCP-OD alignment is described in the remarks area, the Army and FORSCOM designate them under the partnership method" (7). However, this regulation was promulgated eight months after this study began.

CHAPTER TWO

The Division Headquarters: History, Organization, and Roles

This chapter provides necessary context for understanding the challenges of analyzing the readiness or effectiveness of a division HQ design. The first part of the chapter demonstrates that HQ designs have varied throughout recent history for a variety of reasons, while the second part shows that divisions have often been called upon to execute missions for which they were not specifically designed. The chapter concludes with some observations about what makes the FARG II design decisions seem to be outliers in the historical progression of division HQ design.

The Evolution of U.S. Divisions and Division Headquarters

Division HQ have evolved significantly in the American experience from ad hoc and administrative echelons, through formally manned and organized tactical commands, to a HQ capable of both tactical and operational command operating in a joint and multinational environment. Similarly, the structure of the units subordinate to the division have changed from a nonstandardized collection of semi-independent regiments or brigades, to a changing array of structures with formal peacetime command relationships, to the current modular force, with no real formal permanent command or doctrinal relationships. Divisions have also moved from single arms (infantry, cavalry) to combined arms formations, and the number of specialty capabilities that a division commander and staff may have to integrate and command have multiplied—and may be either organic or in support through some other command relationship. As weapons have increased in range, speed, and capability, the span of control for a division HQ has increased across multiple dimensions too. Today divisions plan further into the future, across a larger battlespace, and for more capabilities. Adaptation to each of these changes can be seen in the history of division HQ design.

Early divisional structures tended to be single-purpose or ad hoc organizations that evolved in response to battlefield demands. During the Revolutionary War, Washington's main tactical echelon was the brigade. He established the original American divisions as an administrative echelon above those brigades.[1] Subsequently, the single-arms regiment was the highest-level unit maintained during peacetime, although brigades and higher echelons were created during wartime, particularly during the Civil War. These higher echelons, including divisions and corps, were task organized on an ad hoc basis with minimal staff and no formal staff structure

[1] Robert K. Wright, *The Continental Army*, Washington, D.C.: U.S. Army Center of Military History, 1989, ch. 2.

or doctrinal or standardized subordinate unit structure.[2] At the turn of the twentieth century, Secretary of War Elihu Root sought to standardize the structure and personnel systems of the Army, moving to create an Army that would scale more readily.[3] Simultaneously, the Army's experiences in China and the Philippines led to an awareness that a permanently organized, combined-arms structure was necessary for the wars the United States was likely to fight, and standardized division designs were developed to fill this need.[4] The Field Service Regulations approved in 1905 and updated in 1908 also mandated staff structures and designated divisions as both tactical and administrative units.[5] In 1915, in the wake of the crisis at Vera Cruz, MG William H. Carter recommended the establishment of permanent division HQ to assist in future mobilizations.[6]

Division HQ, just as divisions themselves, changed many times in the century that followed the creation of a permanent staff, following (or responding to) changes in doctrine and warfighting. When divisions deployed to Europe, Pershing formally adopted the numbered staff used by the Europeans, with sections G-1 (personnel), G-2 (intelligence), G-3 (operations), G-4 (supply), and G-5 (training).[7] World War I demonstrated the effectiveness of integrating combined arms, combat support, and combat service support in a single, large tactical unit. But the complex network of relationships through which information, materiel, and soldiers were channeled could not have been sustained or effectively controlled without the division HQ system.[8]

This complex interaction and the demands for synchronization and planning over broader capabilities and time/distance horizons affected division staff size and design. In the first half of the century, new capabilities were added to divisions, and in every redesign after Korea, capabilities and functions were reallocated between echelons (not always in the same direction). In the early years of the Vietnam War, when the 101st Division became airmobile, maintenance was moved from the division level to each individual company, battery, or troop.[9]

Changes in capabilities also affected divisional structures. In the aftermath of World War I, a HQ and HQ battery, commanded by a brigadier general, replaced the field artillery section, in recognition of the crucial nature of coordination at the more rapid tempo of combat between mechanized armies.[10] As aviation and communications technologies developed,

---

[2] Russell Frank Weigley, *History of the United States Army*, New York: Macmillan, 1967.

[3] Walter Kretchik, *US Army Doctrine: From the American Revolution to the War on Terror*, Lawrence: University of Kansas Press, 2012, pp. 108–109.

[4] John B. Wilson, *Maneuver and Firepower: The Evolution of Divisions and Separate Brigades*, Washington, D.C.: U.S. Army Center of Military History, 1998, p. 23.

[5] U. S. Army General Staff, *Field Service Regulations United States Army 1905: With Amendments to 1908*, Washington, D.C.: Government Printing Office, 1908, sec. 1.

[6] Wilson, 1998, p. 35.

[7] Wilson, 1998, p. 67.

[8] Glen R. Hawkins and James Jay Carafano, *Prelude to Army XXI: U.S. Army Division Design Initiatives and Experiments, 1917–1995*, Washington, D.C.: U.S. Army Center of Military History, 1997, p. 6.

[9] Wilson, 1998, pp. 333–334.

[10] Wilson, 1998, p. 144.

intelligence sections grew, and aviation assets and staff were added to division HQ initially.[11] Air and intelligence liaisons were part of division staffs at the end of World War I, and they typically had a single officer in each role, but this manning level grew in later designs.[12] Just as divisions themselves first combined combat arms by incorporating regiments or brigades of different arms, combined capabilities were eventually pushed to ever lower echelons. For example, aviation became an asset of reconnaissance battalions in armored divisions in 1940, even before the United States entered World War II.[13] Over the years, additional capabilities continued to be built in to redesigns of the division, such as the allocation of air defense to division HQ in the 1980s.[14]

HQ staff levels have fluctuated between a low of 102 in the infantry division of 1940 to a high of 670 in the Atomic Field Army division of the early 1950s and 720 for the modular division design of this century.[15] But the size of the divisional HQ has only loosely been related to the size of the division itself. The pentomic division had the smallest number of troops at 8,600 soldiers, but it had one of the largest HQ elements in both relative and absolute terms; the staff, at 566 soldiers, represented 1 person in the HQ for every 15 in the division. The largest division, the square infantry divisions that fought in World War I, had over 27,000 personnel, with 164 on the divisional staff, a ratio of 1 to 165. In general, among Army divisions in the 20th century, infantry divisions up to the end of the World War II had the lowest ratios of HQ personnel to total personnel and smallest HQ in absolute terms; and while numbered staffs were already a feature, each section was smaller than in later infantry divisions or contemporaneous armored divisions.

If staff size was not directly related to the size of the actual division, it does appear to have been related to the level of autonomy expected by subordinate units. In 1942, during World War II, the Army created two (and later three, including a reserve) combat commands for armored divisions, with one of these commanded by a brigadier general and the other two by colonels. While the combat commands owned no assets beyond their HQ, the tank and infantry battalions in the division were allocated to them depending upon the nature of the mission. The concept was to exploit the potential of tanks, with each combat command having infantry and artillery attached tailored to the needs of the particular mission and according to the judgment of the division commander. A 1948 revision placed two of the combat commands (CCA and CCB) each under a brigadier general. This approach allowed greater flexibility in carrying out operations and gave significantly more responsibility to the combat command HQ while reducing that of the division HQ, which now concentrated on configuring and assigning the combat commands.[16] This evolution down in command authority and capability also allowed the division's staff size to shrink.[17]

---

[11] Wilson, 1998, pp. 85–87.

[12] John B. Wilson, "Mobility Versus Firepower: The Post-World War I Infantry Division," *Parameters*, Vol. 13, No. 3, 1983, p. 47.

[13] Wilson, 1998, p. 149.

[14] Wilson, 1998, p. 388.

[15] Richard W. Kedzior, *Evolution and Endurance: The U.S. Army Division in the Twentieth Century*, Santa Monica, Calif.: RAND Corporation, MR-1211-A, 2000, p. 24.

[16] Kedzior 2000, pp. 19–20.

[17] Kedzior, 2000, pp., 19–22.

The magnitude of the HQ of the pentomic divisions of the mid-1950s was largely a function of how and in what conditions they were expected to fight. Placing tactical nuclear weapons under the control of individual battle groups of 1,300 soldiers gave these units more firepower than the 3,600-soldier-strong infantry regiments they replaced. While the pentomic divisions were smaller than ever, though, the five battle groups replaced three regiments, increasing the span of control and thus the demands on the division staff, even though the division was smaller in absolute terms. The doctrine accompanying the pentomic concept also concentrated support and service functions at the division level to an unprecedented degree. Survivability and dispersion meant that battle groups could not include their own tails, as they had in the past; but the chaos of a nuclear battlefield indicated in exercises that echelons above division would be less able to allocate resources, so the division staff also had to absorb those tasks.[18]

Division staff structure and size respond to these changes in command relationships, command roles, and spans of control. In World War I, the span of control for a division was 9 or 10 subordinate units, while in World War II it was 10 to 12.[19] Under the Army of Excellence (AOE) design, the span of control was 13 units. Span of control in the AOE design, despite the proliferation of functions of the division HQ, was kept down by regrouping support functions under a division support command.[20] Under Force XXI, a range of division functions, including fire support, were shifted to the corps level, and improvements in information technology (IT) were used to justify trimming the support element at the division level and reducing the size and, in some senses, the role of the division.[21]

The fungibility of subordinate units of the division compared to the ability to tailor capabilities is a consistent tension in division design that ultimately helped lead to modularity. As has been seen, well before the Army officially became modular, the ability to task organize rather than use an all-purpose structure waxed and waned in division design. While the armored divisions during and shortly after World War II embraced a sort of protomodularity, consideration of how to fight on the nuclear battlefield drove the ROCID (Reorganization of the Current Infantry Division: a variant of the pentomic-era designs) reforms in the opposite direction, with five battle groups, which replaced both the regiment and battalion echelons, designed to be fungible for smooth replacement in the kind of war in which monthly casualties were expected to reach 35 percent.[22]

The modular division HQ, and modularity in general, was a response to three major changes. The first was the wide range of capabilities and operations required within the same conflict, and often the same geography, compared with the major combat operations for which previous force structure and doctrine were designed.[23] The second was a prolonged and low-intensity war, in which units would cycle in and out of combat over a longer period than in the

---

[18] Paul C. Jussel, *Intimidating the World: The United States Atomic Army, 1956–1960*, dissertation, Columbus, Ohio: The Ohio State University, 2004, pp. 78–91.

[19] Wilson, 1998, pp. 39, 41, 165.

[20] Wilson, 1998, pp. 399–400.

[21] Kedzior, 2000, ch. 8.

[22] Jussel, 2004.

[23] Frank G. Hoffman, "Complex Irregular Warfare: The Next Revolution in Military Affairs," *Orbis*, Vol. 50, No. 3, 2006, pp. 395–411.

past.[24] The third was the change in communications and IT that underpinned the so-called Revolution in Military Affairs.[25]

By unencumbering the division HQ from a fixed doctrinal subordinate structure, modularity sought to allow task organization to be an integral part of any deployment and to allow the effort to be sustained over a longer period than if divisions were employed as prestructured units, as in the past. A qualitative shift in the nature of a division HQ also is implicit in modularity. While historically the division has been used as a tactical unit, or unit of action, under modularity the unit of action becomes the brigade, which is supported and directed by the division as the unit of employment.[26] The change has been particularly striking in the latter phases of recent conflicts, when division HQs often deployed for one mission, without any of their home-station brigades, and the brigades deployed to other locations on other time lines. In practice, this means that rather than being the end user of the product of force generation, the division has become part of the process.

Equally important, and as will be discussed in more detail below, the modular division HQ design recognized the increasing role division HQ had been playing as operational-level HQ and in joint and multinational roles. The large staff represented both the lack of permanently subordinate artillery, aviation, intelligence, and other "divisional" units—each of which, in previous designs, had been available to augment the division staff—as well as the need for increased staff capabilities in such areas as civil affairs (CA), military information support operations (MISO), air space management, joint battle space awareness, and multinational coordination.

A range of alternatives to the traditional structure of the division, as the echelon between the brigade and the corps, have been suggested over the past several decades. In the 1990s, the Army considered a "skip echelon" arrangement in which the division and corps would continue to exist as command nodes, but almost all their assets would be attached to brigades or lower.[27] Another attempt to combine rapid communications and decisionmaking with the retention of crucial functions at each echelon is "telescoping," authorizing some communications and data to flow from nonadjacent echelons when appropriate.[28] The very concept of a unit of action and unit of employment arose from Army efforts to determine if or how it could collapse three echelons of command into two (theater or joint task force [JTF], corps, and division) while still providing the capabilities of an in-theater, Title 10–focused, Army component command.

Pushing these changes to a more significant level are the recommendations of COL Douglas Macgregor, who argues that land power dominance is the key to victory and thus deterrence; that changes in technology enable a transformational shift in how the Army fights; and that eliminating the division, and using IT to empower brigades to function with more autonomy, is fundamental to winning in the future. Warfighting functions have been pushed to

---

[24] Les Brownlee and Peter J. Schoomaker, *Serving a Nation at War: A Campaign Quality Army with Joint and Expeditionary Capabilities*, Washington, D.C.: Office of the Under Secretary of the Army, 2004.

[25] Susan F. Bryant, *Forging Campaign Quality: Ensuring Adequate Stability Operations Capability within the Modular Army*, Quantico, Va.: Marine Corps University School of Advanced Warfighting, 2006.

[26] John A. Bonin and Telford E. Crisco Jr., "The Modular Army," *Military Review*, Vol. 84, No. 2, 2004, p. 21.

[27] Peter A. Wilson, John Gordon IV, and David E. Johnson, "An Alternative Future Force: Building a Better Army," *Parameters*, Vol. 33, No. 4, 2003, p. 19.

[28] Francis Fukuyama and Abram Shulsky, "Military Organization in the Information Age: Lessons from the World of Business," in Zalmay M. Khalilzad and John P. White, eds., *Strategic Appraisal: The Changing Role of Information in Warfare*, Washington, D.C.: RAND Corporation, MR-1016-AF, 1999, pp. 327–360.

ever lower levels, he points out, and the organization of the Army must reflect that.[29] As with other attempts to remove responsibilities from one echelon, under Macgregor's plan, both the corps and brigade staffs would be greatly expanded, and the span of control of the corps would increase substantially, but not unmanageably, due to the expanded corps HQ. The Army War College's John Bonin suggests a similar brigade-centric restructure but one in which the corps is diminished in size and role rather than the division.[30]

Previous attempts to collapse echelons have not proven enduring. Before World War II, for example, the brigade was removed from infantry divisions, putting regimental commanders directly under the control of the division HQ. This persisted in the triangular infantry divisions, as well as the armored divisions, and made these regiments more responsive. In 1943, the regimental level was stripped from the armored divisions for all but the three combat commands, further concentrating the resources of the division into combat functions.[31] The brigade was also omitted from the pentomic divisions, out of a need for more rapid response and the desire to compensate for the lethality of tactical nuclear weapons with small, easily dispersible units; under this configuration, the battle group was the only echelon between the division and the battalion. Ultimately, these small groups, without an intermediary echelon, were unable to sustain themselves, and this shortcoming was a factor in the curtailed pentomic experiment.[32]

This summary of division and division HQ design and history demonstrates that division HQ have adjusted as the balance between centralized and decentralized control of assigned units has changed, as the complexity of war has changed, as the temporal and geographic demands on a division have changed, and as the role of the division has evolved from administrative to tactical to operational. But a more mundane influence on the size of division staffs is resource constraints. This phenomenon is noticeable during the world wars, when the cost of manpower in theater meant that staffs were deliberately kept as small as possible, regardless of division function. [33] In the case of the FARG II design, discussed in Chapter Three, resource constraints are for the first time the driving force for reform rather than changes in the nature of combat, anticipated theaters of conflict, or technological development.

The Role of Division Headquarters in Current Army Doctrine and Thought

A division HQ has the following four primary roles in operations:

1. tactical HQ
2. platform for joint or multinational land component HQ
3. platform for JTF HQ in a limited contingency operation
4. Army Forces (ARFOR) HQ for a small contingency.[34]

---

[29] Douglas A. Macgregor, *Breaking the Phalanx: A New Design for Landpower in the 21st Century*, Ex-library edition, Westport, Conn.: Praeger, 1997.

[30] Christopher Kennedy, *The US Army Division: The Continuous Evolution to Remain Relevant*, Carlisle, Pa.: Army War College, 2013.

[31] Kedzior, 2000, ch. 3.

[32] A. J. Bacevich, *The Pentomic Era: The U.S. Army Between Korea and Vietnam*, CreateSpace Independent Publishing Platform, 2012.

[33] Weigley, 1967, p. 424.

[34] Headquarters, Department of the Army, Army Techniques Publication No. 3-91, *Division Operations*, Washington, D.C., October 17, 2014, p. 1-1.

Underlying these different terms is a range of implied variations in what the division HQ must do. The most fundamental can be defined as the "operational compared to tactical" dimension. Historically, divisions were among the highest levels of tactical organizations: they were focused on the immediate engagement with the enemy.[35] On the other hand, according to doctrine,

> The operational level of war links the tactical employment of forces to national and military strategic objectives through the design of campaigns and major operations. It determines how, when, where, and for what purposes commanders employ major forces to achieve assigned ends. It sequences and synchronizes battles, engagements, and other operations (such as disaster relief and support to governance) to achieve operationally significant outcomes. Operational commanders position and maneuver forces to shape conditions for their decisive operation within their assigned operational areas. Commanders exploit tactical victories to gain strategic advantage or reverse the strategic effects of tactical losses. Operational HQ determine objectives and provide resources for tactical operations.[36]

In a prescient 2004 monograph, Kevin Jacobi (then a student at the Army's School of Advanced Military Studies, or SAMS) pointed out that the pattern of using division HQ as the base for JTFs was not just a change in terminology due to the addition of a few non-Army elements. When a Cold War division commander saw the battlefield, his role (they were all "him" at the time) was the senior tactician. The echelons above him were operational, working the lines of logistics, alliances, and so forth to position him to engage and defeat the enemy. He trained with "his" brigades in peacetime, oversaw the deployments of their battalions to the National Training Center or Joint Readiness Training Center, and took them to war. When battle dawned, he would be in his helicopter or tracked vehicle, gazing across the Fulda Gap or the Korean Demilitarized Zone to direct the clash of armor, infantry, and fire support.

Jacobi noted this was not the pattern in the post–Cold War era, even if it took years to see that. Division commanders were leading troops into Mogadishu, Bosnia, Kosovo, or Haiti, not to execute a detailed operational plan from higher but as the senior military leader on the ground. These commanders were dealing with special operations forces, diplomats, aid agencies, partner-nation leaders, the media, and other elements that in the past would have operated in other spheres. The commanders were, in essence, raised to the operational level on the battlefield. At times, they were also tactical leaders, but as often the dispersed forces and mission-command orders made the brigade/BCT commanders the real tacticians, while the generals managed the bigger game.

Wayne Grigsby, who preceded Jacobi at SAMS by several years, focused his student paper on how the HQ's new joint role would have structural implications, increasing both

---

[35] The next higher echelon, the corps, can also be a tactical organization if it is placed under an even higher command in theater, such as a multinational land component command or a U.S. field army. Current Army doctrine notes, "This is the original purpose of the Army corps and the role performed by Army corps in Operation Desert Storm and Operation Iraqi Freedom I. For example, a corps can serves [*sic*] as a tactical land headquarters if war recurred on the Korean peninsula, or if a future crisis led to a general war." Headquarters, Department of the Army, Army Techniques Publication No. 3-92, *Corps Operations*, Washington D.C., April 7, 2016, p. 1-3. Also, see Headquarters, Department of the Army, FM 3-94, *Theater Army, Corps, and Division Operations*, Washington, D.C., April 21, 2014, pp. 4–8.

[36] Headquarters, Department of the Army, ADP 1-01, *Doctrine Primer*, Washington, D.C.. September 2, 2014, pp. 4–10.

the capabilities and the capacity required of the staff.[37] Roughly five years later, George L. Fredrick (also at SAMS) wrote that doctrine for mission-essential task lists (METLs)—upon which units base the focus of their training—was not keeping pace with the changes in the post–Cold War operational environment. In particular, divisions were experiencing security-assistance and stability operations contingency deployments more frequently than major theater war. In addition to Iraq, for example, between 1990 and 2000 the U.S. Army deployed units to "Somalia, Haiti, Macedonia, Croatia, Eastern Slavonia, Hungary, Bosnia, Rwanda, and Kosovo."[38]

Missions and Mission-Essential Tasks of a Division Headquarters

To prepare for the above missions, divisions focus their collective training on the mission of and guidance from the commander of the next higher-level unit and the unit's METL.[39] "A mission-essential task is a collective task on which an organization trains to be proficient in its designed capabilities or assigned mission. A mission-essential task list is a tailored group of mission-essential tasks."[40] For brigade and higher units, the HQ of the Department of the Army (HQDA) standardizes METLs for like-type units. The standardized METL represents the tasks of decisive action that a unit could perform based on its table of organization."[41] Although the current METLs for Army divisions are generally not available to the public, most tasks are drawn from the "Army Universal Task List." These can be found in Field Manual (FM) 7-15, which has been approved for public release and unlimited distribution.[42]

Kevin Jacobi's analysis of the division's operational role, summarized above, was only a prelude to his critique of Army training in the early 2000s. Divisions needed to stop training as if they were tactical HQ, he argued, and focus on what they needed to do as operational commands. The old METLs focused on conducting a defense or movement to contact were misdirected. Having worked in lower-level HQ, staff officers and NCOs should not have trouble serving on a division staff. The challenge would be working at the operational level. But there were no METLs for those tasks.

Reflecting on the 4th Infantry Division's deployment to Iraq in 2007–2008, Alan Batschelet and his coauthors argued that a division's effectiveness is a function of three variables.

1. The division commander's ability to span the tactical operations through strategic conditions over time.
2. The staff's ability to organize and act to create the conditions that lead to realizing the commander's vision.
3. The ability of the division commander and his staff to gain unified action with other agencies and partners as they move toward a common end state.[43]

---

[37] Wayne W. Grigsby Jr., *The Division HQ: Can It Do It All?* Ft. Leavenworth, Kan.: School of Advanced Military Studies, 1996.

[38] George L. Fredrick, *METL Task Selection and the Current Operational Environment*, Ft. Leavenworth, Kan.: School of Advanced Military Studies, 2000, p. i.

[39] Headquarters, Department of the Army, ADRP 7-0, *Training Units and Developing Leaders*, August 2012, p. 3-1.

[40] Headquarters, Department of the Army, FM 7-0, *Train to Win in a Complex World*, Washington, D.C.: October 5, 2016, paragraph 1-41.

[41] Headquarters, Department of the Army, ADRP 7-0, August 2012, p. 3-2

[42] Headquarters, Department of the Army, FM 7-15, *Army Universal Training List*, Washington, D.C.: December 9, 2011.

[43] Alan Batschelet, Mike Runey, and Gregory Meyer Jr., "Breaking Tactical Fixation: The Division's Role," *Military Review*, Vol. 89, No, 6, 2009, p. 35.

These can be seen as the starting point for a reconsideration of METLs as they apply to the operational division HQ.

Looking at division METLs ten years after Jacobi, one finds the same tendencies present today. At both the division and corps level, METLs overwhelmingly focus on the traditional Army tactical model. One challenge this study will attempt to address is how to assess the readiness of division HQ to function in any of their four roles, when most exercises and other efforts focus on one, the tactical HQ.

The Division Headquarters in Operations, 1992–2014

In our initial analysis of the research question, we confirmed that there is usually a significant gap between how the country plans to use the Army (seen in documents like the National Defense Strategy, which focus on critical strategic threats), how division HQ are designed to be employed (the doctrine and structure discussed in the previous sections), and how they are *most likely* to be employed. Because the deployment of a full division HQ and all its assigned subordinate units is relatively rare, we decided to look at a selection of contingency and rotational deployments since 1992 in order to understand the range of the actual division HQ roles and missions. The intent here is not to retell the operational stories; these are well documented in news reports, books, and U.S. Army official histories; instead it is to note the factors that will enable this study to model the implications of MCP-ODs for the division HQ, such as the number of personnel deployed, the duration of the mission, and notice to deploy. The operational deployments selected were the following:[44]

- Operations Restore Hope and Continue Hope, Somalia, 10th Mountain Division (10th Mountain), December 1992–March 1994
- Operation Uphold Democracy, Haiti, 10th Mountain and 25th ID, July 1994–March 1995
- Operation Enduring Freedom, Afghanistan, 10th Mountain, March 2002–April 2005
- Operation Iraqi Freedom, Iraq, 3rd Infantry Division (3rd ID), April 2007–June 2008
- Operation United Assistance, Liberia, 101st Airborne, September 2014–February 2015.

Between 1992 and 2014 the division HQ underwent a transformation in its role. In the Cold War, one could picture a division in the offense role and assume a certain size force, across a certain frontage, against an enemy force with a certain capability. The resulting tasks, their number, and their complexity were well understood. The Army undertook several contingency operations, but these were conducted in a very different and more limited strategic environment. From 1990 onward division HQ tasks and responsibilities have increased dramatically due to the demands of operations. Division HQ have undertaken a range of tasks and missions overseeing forces as small as a single maneuver brigade or as large as multiple BCTs and other elements, engaged across an enlarged area and performing a number of different missions—the "three block war" on the scale of a province or

[44] These are a convenience sample, selected based upon the availability of Open Source data while covering a spectrum of operations that included Humanitarian Assistance/Disaster Relief, stability operations, and major combat operations. Because this is not a random sample, we do not provide a statistical estimate on the representativeness of these cases versus all division deployments from 1992 to 2015 nor on the distribution of mission types.

country.[45] In his 2004 monograph, Jacobi proposed that one of the challenges to training and preparing today's division HQ was the fact that the planned and potential workload is now more varied than ever before.[46]

Taken to its logical conclusion, this study's focus on the multicomponent nature of the HQ may be just the starting point. It may be that the organic, AC manning of the HQ is more than adequate for a range of smaller missions, but if the commander is given three BCTs and a wide range of joint elements, not even the old HQ structure would have been sufficient. The pace of operations is a key variable in the requirements of a HQ. Beyond the size of the divisional force, one needs to know the mix of operational and tactical duties. Is the commander both directing the actions of these assigned units and acting as the senior theater maneuver commander in the operation? If so, the staff's effort will be similarly split between the tactical and the operational, expanding again the workforce required.

Where once divisions operated as the largest tactical formation, increasingly they have taken on operational-level tasks, starting with Operation Restore Hope and Continue Hope in Somalia in 1992–1994 and Operation Uphold Democracy in Haiti in 1994.[47] This has continued due to operational demands in Bosnia, Iraq, Afghanistan, and most recently in Liberia.

Table 2.1 below shows examples of divisional HQ deployments from 1992 to 2016. For the purposes of this study it shows the approximate strength of the deployed HQ (not its doctrinal size) and the duration of the deployment, both factors important for the model. The examples were chosen based on the availability of information from open sources that was verifiable. As such the table is a not comprehensive account of all divisional HQ deployments throughout this period; rather, it is a reasonable selection.

The Concurrent Challenge of Contingencies and Campaigns

From Operation Restore Hope in 1993 through Operation Iraqi Freedom and Operation Enduring Freedom to Operation United Assistance in 2014, the role of the division HQ on operations has been transformed. As discussed earlier, division HQ were once the largest tactical formation but now routinely undertake operational-level roles, as ARFOR, JTF, or combined forces land component command. Whereas in the early 1990s this was exceptional, by the late 2000s the greater span of control, wide geographical dispersion, and sizable political-military challenges became common elements of what the division HQ has to deal with on both rotational and contingency deployments.

The examples in the section also allow for discussion of the variation in notice given to the divisions prior to deployment and the size of the HQ that actually deployed. For example, the 10th Mountain deployed 53 days after notification for its 2.5 month deployment to Operation Uphold Democracy, in 1994. Divisions rotating through Iraq and Afghanistan often had a year to prepare, though the 101st Airborne was given only 29 days to shift its training focus and prepare for a 5-month deployment to Liberia in 2014. In general, it seems that 1

---

[45] Marine Gen. Charles Krulak, widely credited with branding the term, defined it as "contingencies in which Marines may be confronted by the entire spectrum of tactical challenges in the span of a few hours and within the space of three contiguous city blocks." See Charles C. Krulak, "The Strategic Corporal: Leadership in the Three Block War," *Marines Magazine*, January 1999.

[46] K. L. Jacobi, *Division METL—Clinging to an Antiquated Paradigm?* Ft. Leavenworth, Kan.: School of Advanced Military Studies, 2004, pp. 1–2, 21.

[47] Jacobi, 2004, pp. 1–2.

Table 2.1
Sample of Divisional Operational Deployments and Headquarters Size, 1992–2016

Deployment Start	Division	Operation	Duration	HQ Size
September 2016	1st CD	Freedom's Sentinel	12 months	500
March 2016	101st Airborne	Inherent Resolve	9 months	500
November 2016	1st ID	Inherent Resolve	9 months	500
October 2015	10th Mountain	Freedom's Sentinel	9 months	300
June 2015	82nd Airborne	Inherent Resolve	9 months	500
November 2014	3rd ID	Resolute Support	12 months	200
October 2014	101st Airborne	United Assistance	5 months	700
May 2011	1st CD	Enduring Freedom	12 months	850
December 2009	1st ID	Iraqi Freedom	12 months	900
April 2007	3rd ID	Iraqi Freedom	12 months	1,000
July 1994	10th Mountain	Uphold Democracy	6 months	800

SOURCES: Sources for this table include the following, organized by operation: 1st CD (Cavalry Division), September 2016: Donald Korpi, "National Support Element Completes Transfer of Authority from 10th Mountain Infantry Division," Army.mil, September, 13, 2016; Corey Dickstein, "Army: 500 from 1st Cavalry Division Deploy to Afghanistan," *Stars and Stripes*, March 22, 2016. 101st Airborne, March 2016: Nathan Hoskins, "101st Airborne Division Completes Iraq Tour, Transfers Mission to 1st Infantry Division," Army.mil, November 21, 2016; Michelle Tan, "101st Airborne to Deploy to Iraq, Kuwait", ArmyTimes.com, November 6, 2015. 1st ID (1st Infantry Division), November 2016: Army Public Affairs, "Department of the Army Announces 1st Infantry Division Deployment," Army.mil, October 14, 2016; U.S. Central Command, Media Release. 10th Mountain, October 2015: Korpi, September 13, 2016; Kap Kim, "Division Cases Colors for Upcoming Deployment," Army.mil, October 22, 2015. 82nd Airborne, June 2015: Richard Sisk, "82nd Airborne Headquarters Troops to Replace 1st Infantry Division in Iraq," Military.com, April 8, 2015. 3rd ID, November 2014: Michelle Tan, "3rd ID Commander Readies His Troops for Afghanistan," ArmyTimes.com, November 9, 2014; Corey Dickstein, "3rd ID Commander Remaining Focused in Afghanistan Ahead of Command Change," *Savannah Morning News*, July 4, 2015. 101st Airborne, October 2014: James Vidal, "Engineer Battalion Deploys Soldiers in Fight Against Ebola," Army.mil, October 17, 2014; David Vergun, "101st HQ Deploying to Liberia in Response to Ebola Epidemic," Army.mil, September 30, 2014; Chi Truong, "48th CBRN Brigade Uncases Colors in Liberia," *Fort Hood Sentinel*, April 2, 2015; Center for Army Lessons Learned.

to 2 months of notice for a deployment is not an unusual event, or at least less unusual than a no-notice deployment. And again, in general, the deployments are not for major combat operations; nor do they require the full division HQs to deploy. These characteristics create the challenge for the MCU and MCP-OD of being prepared to both meet the needs of sustained campaigns and of being flexible enough to support tailored contingency deployments.

In the next chapter, we describe the processes and decisions that resulted in the current design of division HQ and the creation of the MCP-ODs.

CHAPTER THREE

History of the Focus Area Review Group II Design and Intent/ Limitations

In this chapter, we summarize the genesis of the MCP-OD, a concept that was driven by mandated cuts in the size of corps and division HQ.[1] To appreciate the scale of the changes in the HQ structure, Figure 3.1 shows where the positions in the prior HQ structure went.

Figure 3.2 gives a time line of the key decisions in their development that will be discussed in further detail in the subsequent section.[2]

The 20-percent cut in institutional and operational HQ directed by the Secretary of Defense (SECDEF) was a result of the Budget Control Act of 2011 that significantly reduced defense spending. On August 14, 2013, the CSA directed that a FARG explore reduction of army HQ of 25 percent. In October 2013, HQDA directed Training and Doctrine Command (TRADOC) to redesign division and corps HQ to achieve the directed reductions. The initial

Figure 3.1
Division Headquarters Reductions and Transfers

• Headquarters and headquarters battalion (HHBn) reduced			–25
• Dedicated staff elements eliminated (workload redistributed)		–15	
– Operations research and systems analysis	–2		
– Red team	–3		
– Sensitive compartmented information facility security	–4		
– Mobile Command Group	–6		
• Targeted staff/administration reductions			–37
– Inspector General, chaplain, G3, G2, sustainment, fires	–6		
– Protection, G8, surgeon	–9		
– G6	–22		
• Staff and functions transferred to enabler units		–16	
– Sentinel radar teams and fire support coordination	–3		
– G4 personnel	–13		
• Functions transferred to RC		–128	
– Command liaison officers, public affairs, personnel recovery, Staff Judge Advocate	–17		
– Main command post command operations and information center support elements	–26		
– Fires reductions (staff depth)	–11		
– G1, G2, G3, G4, G8	–60		
– HHBn	–14		

SOURCE: U.S. Army Combined Arms Center, July 25, 2014. (Abbreviations have been spelled out for clarity.)

[1] However, in this report we consider only division MCP-ODs.

[2] U.S. Army Combined Arms Center, 2015, Slides 3–4.

Figure 3.2
Time Line of Division Headquarters Redesign

- July 31, 2013: SECDEF directed 20% cut in institutional and operational HQ August 14, 2013: CSA directive: established FARG to explore reduction of 25%
- August 2013: FORSCOM Commanders Conference for Operational HQ Reduction produced several recommendations:
  - Reduce number of HQ
  - Mission-tailored readiness
- October 2013: HQDA directive:
  - Training and Doctrine Command (TRADOC) (CAC / Mission Command Center of Excellence [MCCOE]) tasked to redesign division and corps HQ
  - MCCOE conducts Council of Colonels
- February 18, 2014: CAC proposed designs to CSA
- July 2014: Design submitted to Army Capabilities Integration Center / HQDA
- November 2014: Corps and division HQ pilot MTOEs approved and documented
- July 2014: BCT reductions directed to reduce Army end strength by FY 2017
- October 24, 2014: MCCOE alerted of additional reductions to division and Corps HQ
- Early November 2014: FORSCOM and TRADOC commanders request additional options for reductions:
  - Reduce division to HQ 350 AC
  - Reduce corps HQ to 450 AC
- Late November 2014: CAC/MCCOE proposal:
  - Division HQ 430 AC
  - Corps HQ 530 AC
- December 16, 2014: CSA decision to not cut division/corps HQ unless required by Budget Control Act
- December 17, 2014: HQDA G-3/5/7 requires cuts by FY 2017, directs a single-component design
- Late January 2015: MCCOE completes "all AC" design:
  - Division HQ 520
  - Corps HQ 570
- February 3, 2015: CSA issues guidance directing multicomponent design (a.k.a. FARG II)

SOURCE: RAND Arroyo Center analysis of U.S. Army Combined Arms Center, December 2015.

result of this TRADOC-led effort was a division HQ reduced from 721 AC soldiers to 481 AC soldiers with the entire division HQ Fires and Air and Missile Defense (AMD) sections (an additional 19 soldiers) residing in the Division Artillery. Some of this lost capacity would be restored through a 128-soldier detachment from the ARNG.[3] In November 2014, HQDA G-3/5/7 approved the authorizations for a pilot effort using this structure to be conducted by the 101st Airborne at Fort Campbell, Kentucky. The 101st Airborne subsequently deployed to Operation Inherent Resolve in Iraq and Kuwait as an MCU, including National Guard soldiers from Utah and Wisconsin.[4]

To prevent additional cuts to other force structure, further cuts were directed to corps and division HQ in what would be termed "FARG II."[5] TRADOC was directed to assess the feasibility of a reduction of the division HQ to 350 AC soldiers. However, based on risk assess-

[3] U.S. Army Combined Arms Center, 2015, Slides 3–4.

[4] Unlike the MCP-ODs developed under FARG II, the National Guard contribution to the 101st Airborne's HQ was composed of individuals moved from their normal units to temporary positions for the purpose of mobilization. Therefore, there was no separate, permanent chain of command for them within the state structure, and the 101st Airborne was a single, multicomponent HQ in Army force-management terms. As we will discuss shortly, this is not a trivial distinction and may have significant impacts on the MCP-ODs.

[5] If one assumes a cap on the size of the Army's active component, then fewer soldiers in division and corps HQ means more soldiers for other units.

ments conducted with division and corps commanders, the division FARG II design ultimately authorized 518 AC soldiers (including the Fires and AMD sections), augmented by 96 soldiers from a MCP-OD, for a small net decrease from the total FARG I HQ. The MCP-ODs are sourced by the ARNG, except in the case of the 25th ID (sourced from the USAR, see Figure 3.3) and a handful of positions on the ARNG documents that are expected to be filled by USAR soldiers. On February 3, 2015, the CSA directed the implementation of FARG II.[6]

Focus Area Review Group II Design Concept

Senior Army leadership directed that FARG II establish multicomponent corps and division headquarters (DIV HQ) capable of fulfilling all doctrinal roles and functions; in other words, the division design had to remain dual capable as a tactical HQ in major combat operations and as a JTF-capable operational HQ or stability operations HQ. The design was to balance risk, in terms of size and capacity, with capability—that is, all essential capabilities were to be retained while managing the risk associated with reduced capacity. For this purpose, TRADOC defined minimal capacity as having at least one soldier with the required skill set. Sections operating at "reduced capacity" might not have the skill present in all duty shifts, while those operating at "full capacity" would have a multishift capability.

Other factors, framed as guidance but which for planning purposes became assumptions, had a significant impact on the ultimate design. In the April 2016 *Division and Corps Reduction (FARG II) Organizational Design Paper*, these were listed as follows:

- The TAC will serve as the primary expeditionary Command Post for Corps and DIV HQs:
  (1) Deployable base of personnel with early entry capability
  (2) Scalable depending on the mission and needs of the commander
  (3) Current operations (OPS) focus; primary command post for future operations
  (4) Composed of AC personnel

Figure 3.3
Focus Area Review Group II Reductions to Division Headquarters

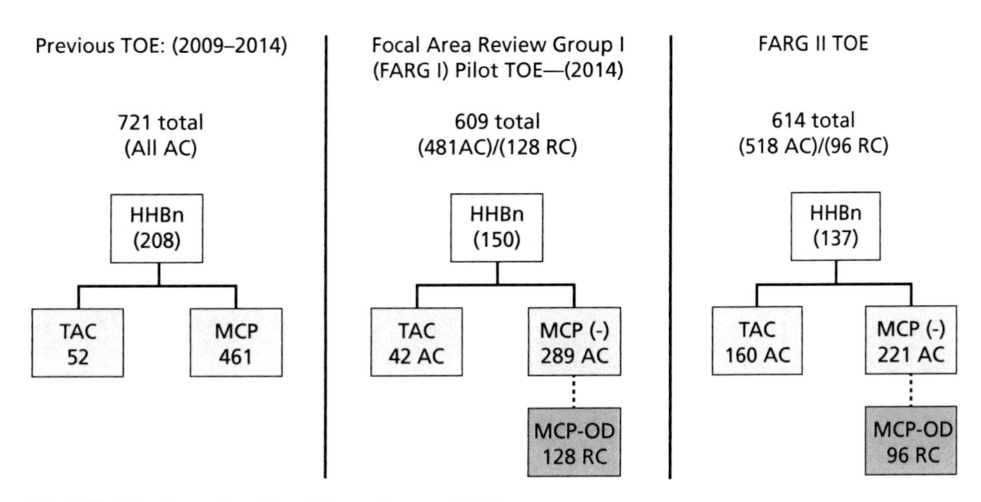

SOURCE: U.S. Army Combined Arms Center, 2015.

---

[6] U.S. Army Combined Arms Center, 2015, Slides 3–4.

- The MCP provides reachback support to the forward deployed TAC from homestation or alternate deployed location:
  (1) If Main deploys into theater, it will be a deliberate action
  (2) Supports the TAC through all operational phases
  (3) Assists with current OPS but focus is on deliberate planning and analysis
  (4) Maintains capability to fully deploy into area of responsibility (AOR) as required
  (5) Composed of AC, RC, and/or Civilian personnel
  (6) RC unit will be stand alone.[7]

This multicomponent design overlaid a fundamental change to the role and size of both the TAC and the MCP. The TAC was designated as the primary expeditionary command post, capable of early entry operations and scalable depending on the mission and needs of the commander. The newly designed TAC was increased from 42 to 160 AC soldiers. Whereas the TAC previously managed only current operations, the new design added the future operations planning horizon. Conversely, the MCP was reduced from 289 to 221 AC soldiers and refocused primarily on division plans. The MCP was redesigned to operate primarily at home station, or in sanctuary, and to provide "reach back" capability for the forward-deployed TAC. Within this division of labor, the MCP-OD was envisioned to provide additional capability and capacity, primarily from home station. However, the MCP is designed to deploy forward, if required. In March 2015, the CSA described the deployment of the MCP as "a deliberate action."[8] This conveys the primacy of the MCP's role from sanctuary, but Army leadership retains the authority to deploy the MCP based on the scale and scope of a contingency. Figure 3.4 depicts the force designers vision of the relationship between the various HQ and command posts during a deployment.

Focus Area Review Group II Risks

The FARG II design assumes risk in terms of the overall capacity of a division HQ. Each billet designated for the MCP-OD represents a requirement (on the TOE) that must be filled for combat operations. By relying on MCP-ODs from the RCs, the design assumes risk in the early periods of an operation, while the RC members are still conducting mobilization and integration activities, which may overlap with key events during the early stages of Phase 3 operations.[9] This window of risk may be further expanded if the MCP is forward deployed.[10]

The design reduces the depth and capacity of the HQ but presumably retains all required capabilities at a level of risk that was deemed by Army leadership to be acceptable. According to several interviewees, necessary capability was attained by maintaining at least one AC soldier for each skill set required in the various sections, cells, boards, and working groups within each

---

[7] U.S. Army Combined Arms Center, *Division and Corps Reduction (FARG II) Organizational Design Paper*, April 15, 2016.

[8] U.S. Army Combined Arms Center, 2015, Slide 5.

[9] Phase 3 operations refer to the "dominate" phase described in joint doctrine. Joint Publication 5-0, *Joint Operation Planning*, August 2011, p. III-39.

[10] Of course, this risk exists in any HQ that is not adequately staffed. The problem is not necessarily limited to MCP-ODs.

Figure 3.4
Focus Area Review Group II Headquarters Employment Concept

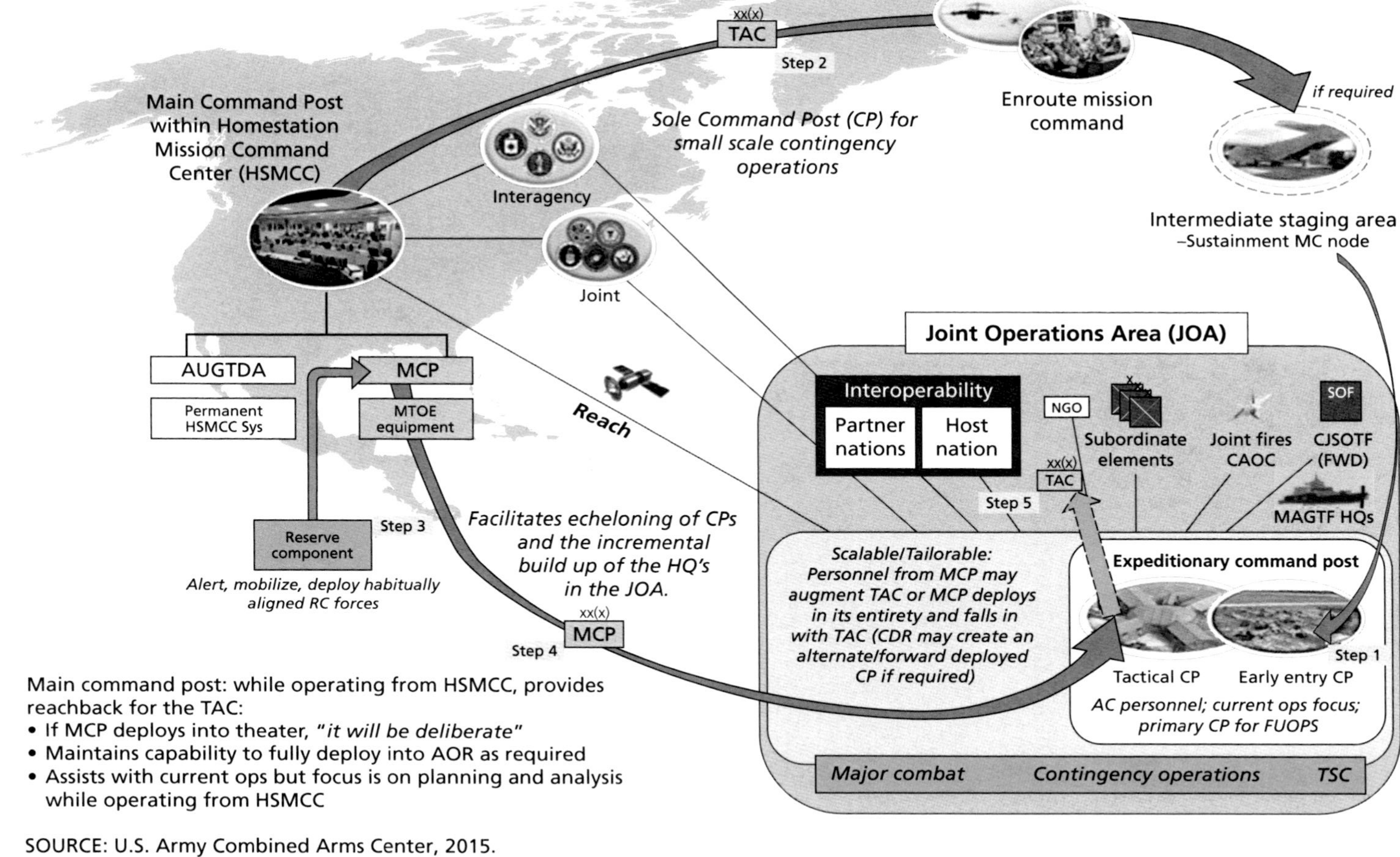

SOURCE: U.S. Army Combined Arms Center, 2015.
KEY: AUGTDA = augmentation table of distribution and allowances; CAOC = Combined Air Operations Center; CJSOTF = Combined Joint Special Operations Task Force; MAGTF = Marine Air/Ground Task Force; NGO = nongovernmental organization; TSC = theater security cooperation.

command post.[11] Capacity was measured by how many shifts contained the required skill sets. Thus, if a section, cell, board, or working group had two or more soldiers with similar skill sets, authorized positions could be either cut or moved to the MCP-OD MTOE. Absent a MCP-OD, soldiers in the remaining authorized positions would be either expected to cover multiple shifts, or some shifts would go without those capabilities, and some tasks would have to be performed sequentially rather than simultaneously. However, interviewees also stated that no formal workload analysis was conducted. The criterion was binary: either a section, cell, board, or working group had the capability available, or it did not.[12]

A force design team at the CAC summarized the known design risks as follows:

- Risk is to depth and capacity rather than capability; increases risk to resiliency in extended/prolonged operations
- Simultaneous execution of installation senior commander responsibilities, Training Readiness Authority oversight and operational deployments may overstretch capacity
- HQ personnel available for daily operations were reduced at Corps by 190 and by 203 at division from the prior MTOE
- Less capability to make a rapid transition to JTF (operating up to 120 days without Joint Manning Document support)
- Intelligence, with the elimination of 9 x 35F (Intel Analyst) personnel at Corps and movement of an additional 7 x 35F to RC and movement of 13 x 35F personnel to RC at division
- Personnel and property management with reduction of HHBn from 4 to 2 companies
- Increases dependency on enablers with a need for habitual alignment of some enabler capabilities (Expeditionary Military Intelligence Brigade, Maneuver Enhancement Brigade [MEB], Sustainment Brigade)
- Reachback concept increases dependency on satellite and network communications which could be exploited by an enemy with antisatellite capabilities.[13]

The force design team further divided the above list into risks specific to division roles and functions. "Synchronize Joint and Army capabilities" and "JTF and JFLCC [joint force land component commander] for limited contingencies" were categorized as medium risks; they categorized as "Low Risk" all of the following:

- Tactical HQ
- Translate major operations plans into tactical actions
- Task organize brigades and battalions

---

[11] Sections are found in the formal structure of the command post, while a cell is "a grouping of personnel and equipment organized by warfighting function or by planning horizon," and "commanders establish boards, working groups, and planning teams to coordinate action and solve problems." Headquarters, Department of the Army, Army Training Publication 6-0.5, Command Post Organization and Operations, March 2017, pp. 2–4 and 2–8, respectively.

[12] Six CAC AC officers and Department of the Army civilians, Ft. Leavenworth, Kan., January 4, 2017; Doctrine writer who was on FARG II design team and an operations research analyst, January 4, 2017, Ft. Leavenworth, Kan.; Four directors, deputy directors, and branch chiefs at the CAC who had been involved with FARG design and implementation, January 5, 2017, Ft. Leavenworth, Kan.

[13] U.S. Army Combined Arms Center, 2015, Slide 9. "Without Joint Manning Document support" means the length of time the division HQ would operate before a Joint Manning Document for the specific mission could be developed, approved, and filled by the services.

- Employ BCTs and other brigades in combined arms operations and
- ARFOR within JTF.[14]

Many division commanders serve as the senior mission commanders for their installations. FARG II reductions drastically reduced the capacity to execute Training Readiness Authority while simultaneously providing oversight to operational deployments. This includes a reduction of 203 active duty soldiers from the previous design.[15] Thus, division commanders require additional capacity to fulfill Title 10 responsibilities[16] on their installations at a time when the tables of distribution and allowances (TDAs) authorizations for civilians who augment uniformed personnel have been drastically reduced. It is important to note that the division HQ is organized for Phase 3 operations and not Title 10 support. A typical division HQ will task organize to meet certain Title 10 roles for which it is not designed or resourced. For example, a training cell is not authorized, so division HQ typically builds one on an ad hoc basis at the expense of other sections. The civilian TDAs are designed specifically for Title 10 functions, to include the execution of Training Readiness Authority, but civilian workforce reductions and hiring freezes have limited their ability to augment the division. By reducing the division HQ by 203 soldiers, to include the 96 soldiers sourced by the MCP-OD, FARG II compounds the existing challenge of carrying out these Title 10 roles. In other words, the reduced capacity severely limits the flexibility to provide necessary capability for missions required at home station but which are not reflected in a go-to-war MTOE. While this effect on the burden carried by the HQ is not included in this project's model of divisional readiness, it clearly will have an indirect impact and may merit more specific analysis and mitigation strategies.

The process of reshaping the division HQ did seem to consider overall peacetime workload in deciding how much to reduce each section. Figure 3.5 takes the composition of the new HQ shown in Figure 3.3 down a layer and shows how the 96 RC positions are distributed among the staff sections. Compared to other sections, FARG II disproportionately cut the intelligence capability of the division. The G2 section was reduced by 37 AC soldiers, including 32 positions to be sourced on mobilization by the MCP-OD. The MCP-OD positions in the G2 affect significant intelligence function to include 14 all-source intelligence analysts (35F MOS series); 7 imagery analysts (6 35G MOS series, and 1 12Y geospatial engineer); 5 signals intelligence analysts (35N); 3 human intelligence collectors (35M); 2 operations officers (35D); and 1 military intelligence system maintainer/integrator (35T). These reductions reflect the assessment that operations in garrison may not require the large staffs historically

[14] U.S. Army Combined Arms Center, 2015, Slide 10. The slides do not give details on how the risk assessments were made, other than that they reflect the input from staffing the FARG II changes with corps and division commanders.

[15] The 101st Airborne was the only HQ to transition to the FARG I design. All others transitioned to the FARG II design, which authorizes 518 AC soldiers, from a previous design of 721 authorizations—a net loss of 203 soldiers. These numbers are based on the approved Force Design Update brief. See U.S. Army Combined Arms Center, 2015.

[16] "Title 10 responsibilities" is military shorthand for the tasks undertaken to sustain military forces, generally a service responsibility, as opposed to the operational command responsibilities generally exercised by combatant commanders working for the President through the Secretary of Defense. The Secretary of the Army's enumerated Title 10 responsibilities and authorities include the following functions: "Recruiting, organizing, supplying, equipping (including research and development), training, servicing, mobilizing, demobilizing, administering (including the morale and welfare of personnel), maintaining, construction, outfitting, and repair of military equipment, and the construction, maintenance, and repair of buildings, structures, and utilities and the acquisition of real property and interests in real property." U.S. Code, Title 10, Subtitle B, Part I, Chapter 303, §3013b. Many of these are passed down the chain of command to the senior commanders on Army installations.

Figure 3.5
Divisional MCP-OD Positions by Section

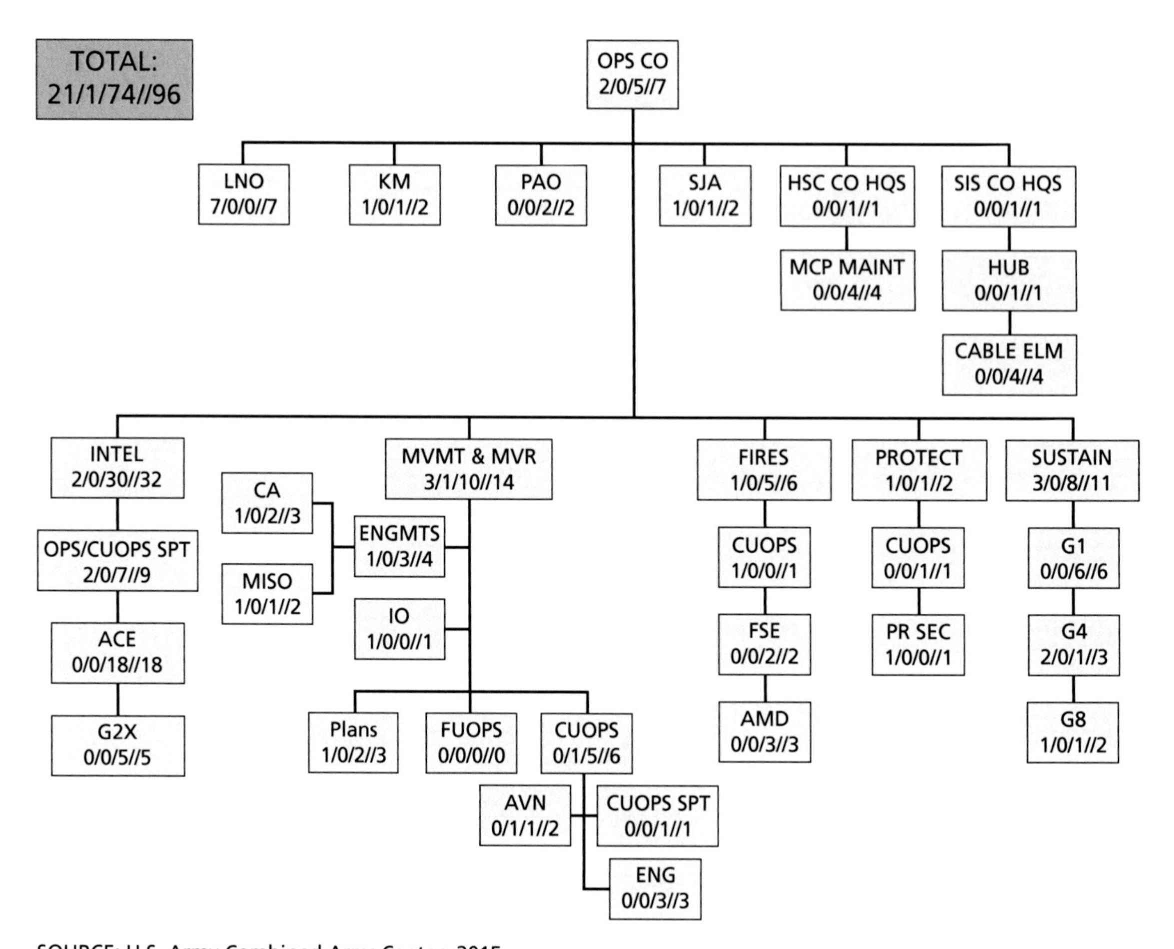

SOURCE: U.S. Army Combined Arms Center, 2015.
NOTE: Figure 3.4 depicts the sections within an MCP-OD.
KEY: Officer/Warrant/Enlisted//Total; ACE = all-source collection element; AMD = air and missile defense; AVN = aviation; CO = company; CUOPS = current operations; ELM = element; ENG = engineer; FSE = fire support element; FUOPS = future operations; HSC = headquarters and supply company; IO = information operations; KM = knowledge management; LNO = liaison officer; MAINT = maintenance; PAO = public affairs office; PR SEC = personnel recovery section; SIS = signals, intelligence, and sustainment; SJA = staff judge advocate; SPT= support.

associated with the G2 section. By outsourcing key intelligence positions to the MCP-OD, the FARG II design assumes risk in the ability of the institutional force to provide these billets that entail highly technical skills with long lead times for initial entry training and certification.[17] This requirement can challenge the ability of the ARNG to deliver these positions, particularly in states that do not have a large supply of intelligence units. (This is not meant to suggest that it is inherently hard to identify and train an individual military intelligence [MI] soldier, but the conventional wisdom holds that filling a number of positions of any type is easier if there are other units in the region with similar positions at lower or equivalent grades or civilian positions producing workers.) Figure 3.6 goes yet another layer deeper to depict the AC-RC breakdown within the G2.

[17] For example, the Intelligence Analyst Course is 16.5 weeks of training, the Imagery Analysts Course is 21 weeks of training, and the Signals Intelligence Course is 25 weeks of training. See U.S. Army, *Military Occupation and Classification Structure*, Department of the Army Pamphlet, 611-21, Washington, D.C.: U.S. Army, August 10, 2008.

Figure 3.6
Active Component and Reserve Component Positions Within the G2

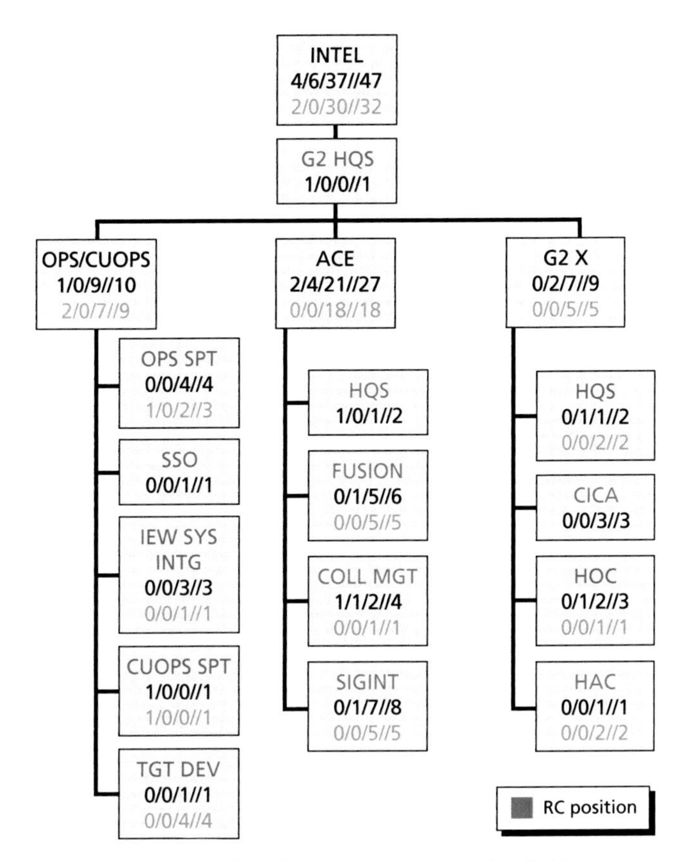

SOURCE: U.S. Army Combined Arms Center, December 2015.
KEY: CICA = counterintelligence coordinating authority; COLL MGT = collection management; CUOPS = current operations; HAC = human intelligence analysis cell; HOC = human intelligence operations cell; IEW = intelligence and electronic warfare; SIGINT = signals intelligence; SPT = support; SSO = Special Security Office; TGT DEV = target development.

The new HQ design also relies heavily on the MCP-OD in the movement (MVMT) and maneuver (MVR) and sustainment (SUS) warfighting functions. The current operations cell has six MCP-OD positions: two aviation positions, three engineers, and one in the current operations integration center. The engineer positions include senior noncommissioned officer positions, thus creating a more rank-heavy section.[18] Reductions to the AC enlisted population are particularly sensitive because labor-intensive tasks, such as vehicle maintenance and command post setup, may fall on a decreasing number of soldiers and noncommissioned officers. The sustainment enterprise also relies heavily on the MCP-OD with 11 positions filled by RC personnel. The G-1 has 6 positions coded MCP-OD, the G-4 section has 3, and the G-8 has 2. Based on overall FARG II reductions, the division has increased dependencies on enabler units, such as the Expeditionary Military Intelligence Brigade, MEB, and the Sustainment Brigade.[19]

---

[18] Engineer section MCP-OD positions include one senior engineer NCO, E-8, 12Z50; one operations sergeant, E-7, 12B40; one operations sergeant, E-6, 12B30. It should be noted that documented MTOE authorizations vary from initial Force Design Update. Source: FMSWeb.

[19] U.S. Army Combined Arms Center, 2015, Slide 9.

Figure 3.7
Focus Area Review Group II Division Main Command Post

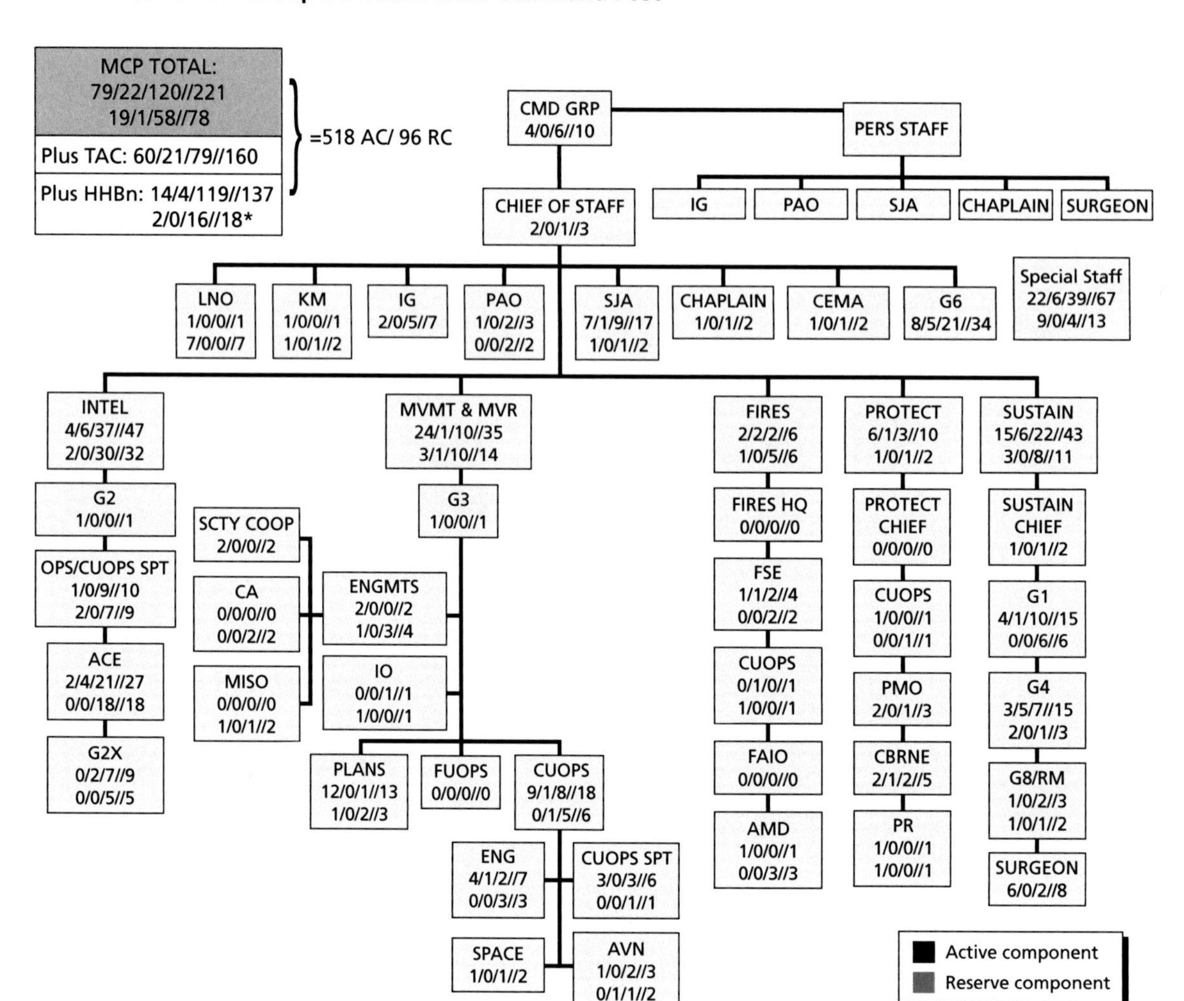

SOURCE: U.S. Army Combined Arms Center, 2015.
NOTE: Includes 7 personnel from MCP-CO Company Staff.
KEY: CBRNE = chemical, biological, radiological, nuclear, and high-yield explosives; CEMA = cyber electromagnetic activity; FAIO = field artillery intelligence officer; FSE = fire support element; IG = Inspector General; MVR = maneuver; SCTY COOP = security cooperation.

Figure 3.7 depicts the MCP authorized personnel at the reduced FARG II level. Figure 3.8 depicts FARG II division tactical command post (DTAC) authorized personnel, which is all AC personnel in its current construct and therefore mostly relevant here to show the overall distribution of functions between command posts.

"Fielding" Schedule of Focus Area Review Group II Divisions and of Main Command Post–Operational Detachments

Following the decisions to create the MCP-OD and in designing its structure, the Army made several secondary decisions about how it would be fielded and employed. These included the following:

- Locations: which states would provide MCP-ODs, and where the MCP-ODs would be located

Figure 3.8
Focus Area Review Group II Division Tactical Command Post

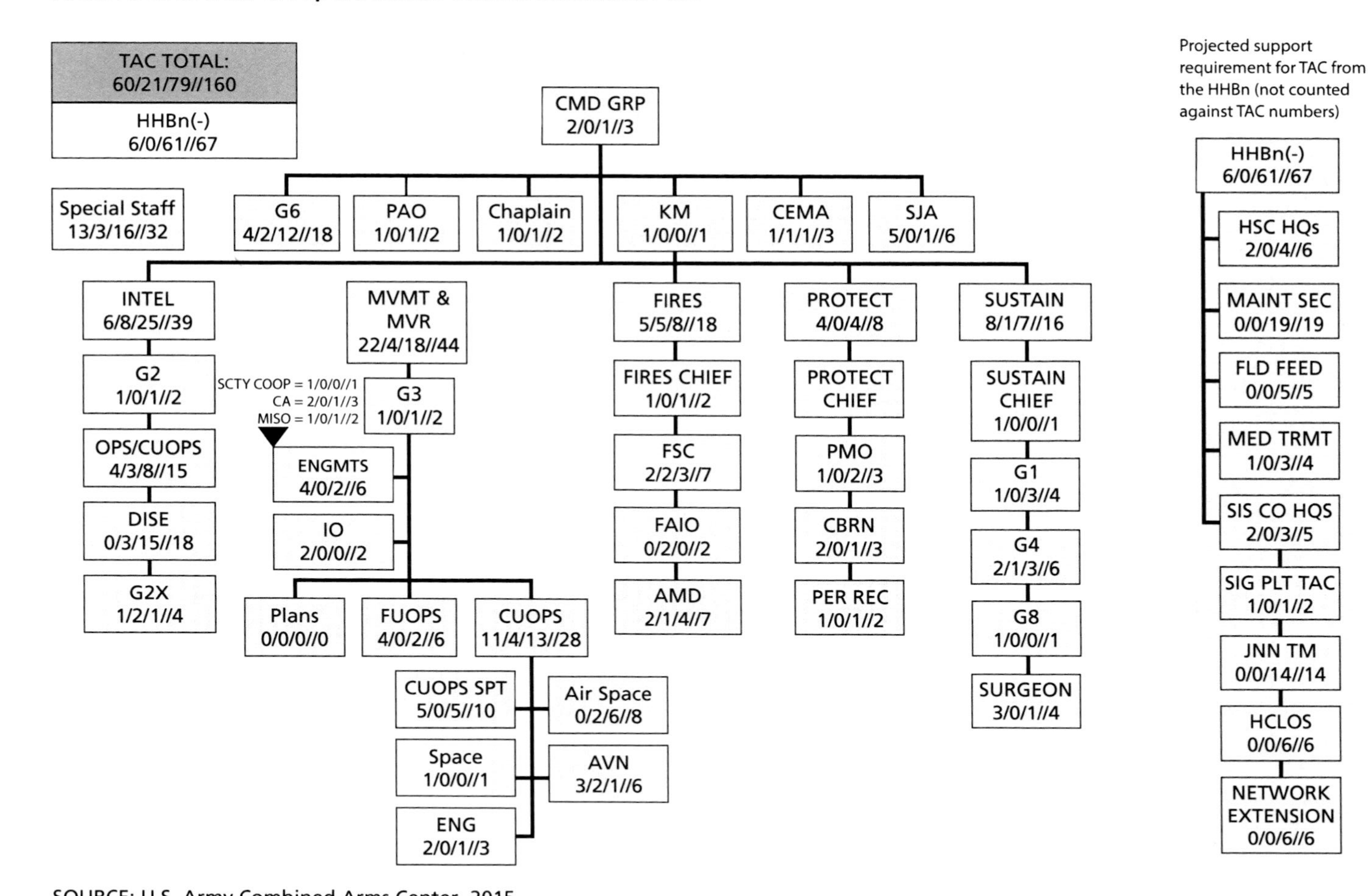

SOURCE: U.S. Army Combined Arms Center, 2015.
KEY: CBRN = chemical, biological, radiological, nuclear; CMD GRP = command group; DISE = deployable intelligence support element; FLD FEED = field feeding; FSC = forward support company; JNN TM = Joint Network Node Team; PLT = Platoon; PMO = Provost Marshall Office; SIS CO = Signals, Intelligence, and Sustainment Company; SIG = Signals.

- Command relationship: whether MCP-ODs would be under the direct command and control of the AC division commander at all times or would have some less direct supporting relationship
- Readiness reporting: would the division commander include readiness of the MCP-OD in assessing the readiness of his HQ to perform mission-essential tasks, or would the two units report readiness as if they were independent units?

These issues and other detail questions of how the MCP-ODs are being fielded will be addressed in the most relevant DOTMLPF-P sections in Chapter Four.

CHAPTER FOUR

DOTMLPF-P Analysis of Focus Area Review Group II Impacts

Introduction

In this chapter, we use the DOTMLPF-P framework to organize our observations. In most cases an observation relates to multiple domains within this framework, but we have chosen to list it only once according to what we assessed to be the most pertinent aspect. Most sections are subdivided into particular issues in that function, and within each of those issues we discuss its source, how we attempted to measure or assess it, and the possible ways to mitigate any negative effects it may have on the consolidated division HQ's readiness to deploy.

Doctrine and Policy

According to JP 1, "Joint doctrine consists of the fundamental principles that guide the employment of US military forces in coordinated action toward a common objective. It provides the authoritative guidance from which joint operations are planned and executed."[1] Within the context of this study, doctrine describes how divisions conduct combined arms maneuvers, perform wide area security operations, or carry out stability operations. The Army's doctrine for the division can be found in FM 3-94, *Theater Army, Corps, and Division Operations*.[2] It explains how the division HQ is organized and describes its command posts. Additional guidance is provided in Army Techniques Publication (ATP), No. 3-91, *Division Operations*.[3] ATP No. 6-0.5, *Command Post Organization and Operations*, also describes a range of command post types, their organization, and their functions.[4]

From a force management perspective, "doctrine analysis examines the way the military fights its conflicts with emphases on maneuver warfare and combined air-ground campaigns to see if there is a better way that might solve a capability gap. Is there existing doctrine that

---

[1] Joint Chiefs of Staff, JP 1, *Doctrine for the Armed Forces of the United States*, Washington D.C., March 25, 2013, p. xxiv.

[2] Headquarters, Department of the Army, FM 3-94, 2014.

[3] Headquarters, Department of the Army, October 17, 2014.

[4] Headquarters, Department of the Army, Army Techniques Publication No. 6-0.5, *Command Post Organization and Operations*, Washington, D.C., March 1, 2017.

addresses or relates to the business need? Is it Joint? Service? Agency? Are there operating procedures in place that are NOT being followed which contribute to the identified need?"[5]

Policy concerns the set of "Department of Defense [DoD], interagency, or international policy issues that may prevent effective implementation of changes in the other seven DOTMLPF-P elemental areas."[6] Especially when it comes to direction from such documents as high-level command training guidance or the direction from the CSA for FARG II to incorporate a multicomponent design, it is sometimes difficult to distinguish between "doctrine" and "policy." Because doing so would not contribute to this analysis, we broadly address both constructs in this section without following a strict typology.

We identified four major doctrinal and policy implications of FARG II: (1) division command post operations, (2) the concept for employment of the division main command post, (3) time lines of MCP-OD availability, and (4) readiness reporting. These implications are interrelated.

Division Command Post Operations

FM 3-94 states that the division HQ "consists of a main command post, tactical command post (which may be reorganized into an EECP [early entry command post]), mobile command group, and HHBn. Plans and operations across these staff sections, cells, command posts, and echelons are facilitated by a network and suite of mission command systems to enhance collaboration and synchronization."[7] The nomenclature "MCP-OD" and the position titles in its modified TOE imply that the soldiers assigned to this unit will primarily perform their duties in the MCP. However, during the WFX that we observed it was common to find MCP-OD soldiers also in the DTAC and support area command post (SACP).[8]

Such assignments in themselves are not necessarily contrary to doctrine. As ATP 3-91 states: "The division commander has the flexibility to organize the five components of the division's mission command system to support that individual's ability to make decisions and facilitate communication within the division HQ as well as with higher, subordinate, adjacent, and supporting commanders."[9]

What such employment outside the MCP illustrates, however, is that (1) practice appears to differ from what the designers of FARG II anticipated and (2) practice is arguably getting ahead of current doctrine.[10] All of the divisions we observed had established or adopted an additional (i.e., third) division command post for their WFX.[11] The majority

---

[5] AcqNotes, "JCIDS Process: DOTMLPF-P Analysis."

[6] AcqNotes, "JCIDS Process: DOTMLPF-P Analysis."

[7] Headquarters, Department of the Army, FM 3-94, 2014, pp. 6–13.

[8] In a few cases, interviewees referred to a "security area command post" that was colocated with the division's MEB. However, we assessed that this simply indicated confusion about how to spell out the acronym "SACP" rather than a different type of command post.

[9] Headquarters, Department of the Army, 2014, p. 2-2.

[10] Practice getting ahead of doctrine is neither unusual nor necessarily bad. It does, however, add a layer of complexity when assessing the impact of MTOE changes.

[11] We did not observe the April 2016 WFX of the 1st Infantry Division (1st ID) but note that an article written by two staff members stated that the division "pioneered the use of a support-area command post (SACP) to command and control the rear area . . . [which was] under the command and control of a deputy commanding general." The article also reports that the division "further integrated critical staff members from its newly established 1ID Main Command Post Operational

of interviewees described it as a "SACP" but a few used the term "DREAR" (division rear command post).[12]

Whether called SACP or DREAR, we observed the following characteristics of these command posts:

- MCP-OD soldiers were present in most cases.
- They have broader responsibilities than the MEB's MCP described in FM 3-81.[13]
- They often have a significant presence from the division HHBn in addition to the MEB's HQ and HQ company (HHC).
- In several cases, they appeared to be the primary duty location of the division's deputy commanding general-support.

As one senior officer described it, "It is not current doctrine, but we have basically reincarnated the old DREAR concept."[14] However, according to FM 3-81, *Maneuver Enhancement Brigade*, "The MEB HQ may be used as an additional division CP or to reinforce one."[15]

The question of whether our observations indicate an expansion of the responsibilities of the SACP or the establishment of a nondoctrinal DREAR is beyond the scope of this study.[16] The key point is that we routinely observed MCP-OD soldiers whose duty locations were not inside the MCP. Especially in cases where MCP-OD soldiers are employed in the DTAC, these observations have implications for how quickly such command posts can be deployed without substantial warning and may also undermine the validity of some of the assumptions on which FARG II was based. Furthermore, if some degree of ad hoc organizing—reflecting flexibility and agility—of command post structure is typical, then assessing the impact of MCP-OD readiness is made more difficult.

Current doctrine considers echeloning division command posts in terms of battlespace. Consideration of echeloning in terms of time may require changes to doctrine or new Army techniques to account for some parts of a command post (i.e., the slots with AC soldiers) being available sooner than others (i.e., slots occupied by RC soldiers). It may be that commanders need to plan for a command post to build up over time, just as they do with BCTs and other elements that are limited by shipping and airflow capacity.

Concept for Employment of the Main Command Post

ATP No. 6-0.5, *Command Post Organization and Operations*, provides just a hint of the assumptions underlying FARG II. Appendix E, "Division and Corps Redesign," summarizes a "new

---

Detachment from the Nebraska National Guard." Jerem G. Swenddal and Stacy L. Moore, "From Riley to Baku: How an Opportunistic Unit Broke the Crucible," *Military Review*, January–February 2017, pp. 77, 82.

[12] See Headquarters, Department of the Army, FM 71-100, *Division Operations*, Washington, D.C., August 28, 1996, ch. 3.

[13] Headquarters, Department of the Army, Field Manual 3-81, *Maneuver Enhancement Brigade*, Washington, D.C., April 21, 2014, pp. 1-4–1-6.

[14] AC field-grade officer, Schofield Barracks, April 8, 2017.

[15] Headquarters Department of the Army, Field Manual 3-81, 2014, p. 1-14.

[16] RAND has a separate, ongoing research project on the implications of the SACP for division operations and the potential shortfall in active duty MEBs.

headquarters design" that includes "a home station CP and a forward CP." The stated purpose for restructuring the HQ into these two command posts is "to reduce deployment time and increase the mobility, agility, and survivability of a headquarters."[17]

The FARG II "Force Design Update Brief" referenced above indicates an expectation that the MCP-OD personnel would typically operate in the home station command post while the forward command post or DTAC would be staffed only with AC soldiers. However, none of the WFX we observed seemed to operate under a home station command post condition. In each of these scenarios, the full division HQ had deployed to the area of operations.[18]

A complicating factor, reported by multiple interviewees, was that command posts during the WFX we observed tended to be overstrength due to personnel assigned in excess of MTOE authorizations or augmentation from subordinate units or other sources.[19] Interviewees involved in the FARG II design and subsequent mission command training stated that the design assumed that additional capacity could be obtained as necessary by attachments, or "plugs," aside from the MCP-OD. For example, public affairs officers in the division HQ could be obtained from public affairs detachments attached to the division.[20] One senior staff officer stated that if the division had a Global Response Force (GRF) mission and the MCP-OD would not be available for 60 days, "we'd rob a brigade."[21] While divisions probably look first to subordinate units, they can also request support from their higher HQ, FORSCOM, or the rest of the Army, depending on the required skills and grades.

Perceptions about the impact of the MCP-OD being present or not present may therefore have been skewed by the context of the exercise. Additionally, because command posts operating in the field are organized by functional and integrating cells—as illustrated below in Figure 4.1—rather than MTOE lines and paragraphs, it is not immediately obvious which MTOE positions are aligned with what command post functions.

As one senior NCO stated: "The MCP is 'battle staff' organized differently from what is shown on the MTOE. It is centered on IRCs [interrelated capabilities]."[22] None of our interviewees stated that they were aware of a crosswalk between MCP-OD positions and specific duties or duty locations within a command post. One officer reported that the division chief of staff had planned to create one but was now waiting for the command post standard operating procedures (SOPs) to be revised after the WFX.[23] A frequent comment from AC interviewees was that unless one happened to know that a soldier was assigned to the MCP-OD (or otherwise from the RC), they were indistinguishable from AC soldiers inside the com-

---

[17] Headquarters, Department of the Army, 2017, p. E-1.

[18] The distinction between short- or no-notice contingency operations compared with rotational conditions is explored in greater detail in the next chapter.

[19] Division primary staff officer, Schofield Barracks, Hawaii, April 8, 2017; Division operations section officer, Ft. Bragg, N.C., June 14, 2017; Division intelligence section officer, Ft. Bragg, N.C., June 14, 2017; Three NCOs (E-7 thru E-9) in SACP, Ft. Bragg, N.C., June 13, 2017.

[20] Directors, deputy directors, and branch chiefs at the CAC who had been involved with FARG design and implementation, January 5, 2017, Ft. Leavenworth, Kan.

[21] AC senior field-grade officer, January 10, 2017, Ft. Campbell, Ky.

[22] Division functional cell NCOIC, Schofield Barracks, Hawaii, April 9, 2017.

[23] Operations section officer, Schofield Barracks, Hawaii, April 10, 2017.

Figure 4.1
Command Post Functional and Integrating Cells

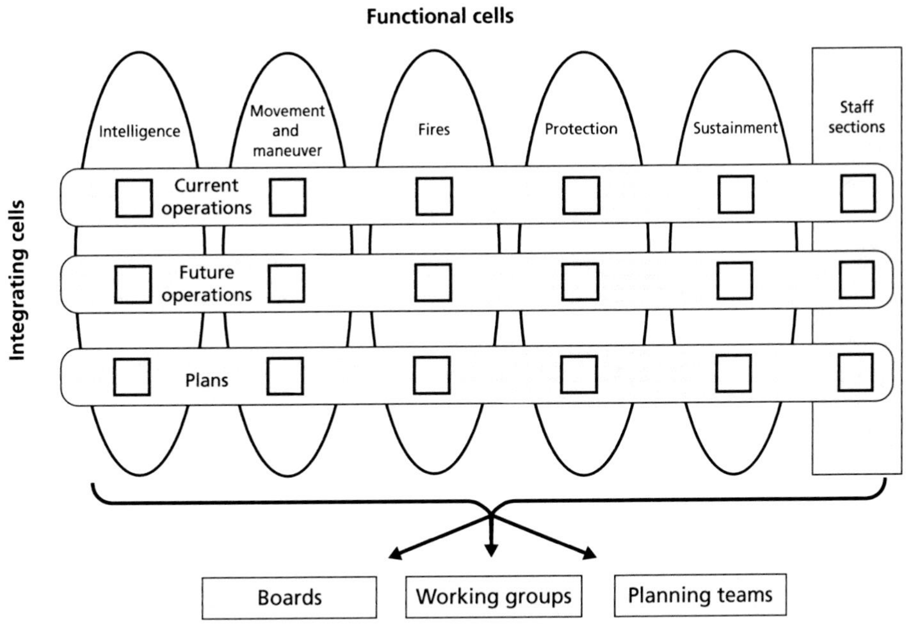

SOURCE: Headquarters, Department of the Army, March 1, 2017, p. 2-8, Figure 2-3, "Cross Functional Staff Integration."

mand posts. According to one field grade officer: "Most of the AC soldiers cannot tell who is a MCP-OD soldier versus an AC soldier."[24]

The bottom line for these observations is that MCP-OD soldiers seem likely to fit into the command post in the same way augmentees generally do, by showing up and plugging a hole, not by reporting to a seat planned in advance and reserved for them. At that point, it is up to them and their new section, cell, or team to bring them up to speed.

Time Lines and Frequency of Main Command Post–Operational Detachments Availability

FORSCOM mobilization policies for MCP-ODs appear to only anticipate their use in operations with relatively long planning time frames. (These are commonly called "patch chart" deployments.) Division commanders who wish to access all or part of their partnered MCP-OD must first perform a mission analysis that considers the following factors:

- Authority: Is there a valid mobilization authority that can be used to support the mission?
- Time: Is there sufficient time for an RC unit to prepare for the mission that meets policy and statutory requirements? If not, is a waiver to policy/statutory requirements justified?
- Funding: Is there funding available for RC pay and allowances?
- Mobilization to dwell (MOB to Dwell): Does the requested MCP-OD have the required one-to-four MOB to Dwell threshold that would allow this unit to be mobilized based on its current sustainable readiness time line?

[24] Division operations section officer, Ft. Bragg, N.C., June 14, 2017; AC senior field-grade officer, Ft. Campbell, Ky., January 10, 2017; Three AC officers, Ft. Campbell, Ky., January 9, 2017; AC sergeant major, Ft. Bragg, N.C., June 14, 2017.

If the mission analysis determines that MCP-OD deployment is required, the request must be submitted to Commanding General of FORSCOM, who submits the request for mobilization through channels for final approval by the SECDEF for MCP-OD mobilization, at least 270 days prior to the date that mobilization of the MCP-OD is required. Furthermore, MCP-ODs do not receive any additional training days beyond the standard 24 inactive duty training (IDT) days and 15 annual training (AT) days each fiscal year until they are officially notified that they will be mobilized.[25]

Several interviewees stated that the original FARG II concept was based upon MCP-ODs routinely having additional training days every year instead of only after notification of sourcing.[26] It further appears that the original concept was to make the division HQ a multicomponent unit that included the MCP-OD personnel on the same authorization document. Establishing the MCP-OD as a multicomponent unit, which FORSCOM describes as the "most structured method for integrating AC and RC organizations," might have mitigated some of the residual challenges we found related to readiness reporting, synchronization of training, and sourcing of CA and MISO personnel. However, the MCP-ODs were instead designated partnership units, the relationship that is "the least structured and most flexible method for integrating units."[27]

The consensus regarding the experience to date with MCP-OD deployments, including the similar but pre–FARG II MCU deployment with the 101st Airborne and the 1st ID and 3rd ID, is that the concept has been successful in providing trained and ready personnel *given sufficient notification and additional training time* to allow the state to fill positions and complete required training.[28]

However, the driving theme in the FORSCOM Command Training Guidance for FY 2018 is "Ready Now." Commanding GEN (CG) Robert Abrams has stated that the Army Forces General Model "must be replaced in practice and mindset by Sustainable Readiness." Reserve component commanders are told "to prioritize Soldier personnel readiness over collective training requirements during pre-mobilization periods."[29] The guidance from the FORSCOM CG for units appears to conflict with that of the CSA in accepted FARG II

---

[25] Email from FORSCOM mobilization planner, July 18, 2017. However, several interviewees stated that the original FARG II concept was based upon MCP-ODs routinely having additional training days every year instead of only after notification of sourcing. (Six Combined Arms Center AC officers and Department of the Army Civilians, Ft. Leavenworth, Kan., January 4, 2017; Four directors, deputy directors, and branch chiefs at the Combined Arms Center who had been involved with FARG design and implementation, January 5, 2017, Ft. Leavenworth, Kan.) Also, see U.S. Army Combined Arms Center, 2015.

[26] Six Combined Arms Center AC officers and Department of the Army Civilians, Ft. Leavenworth, Kan., January 4, 2017; Four directors, deputy directors, and branch chiefs at the Combined Arms Center who had been involved with FARG design and implementation, January 5, 2017, Ft. Leavenworth, Kan.

[27] For a full description of partnership units and MCU arrangements, see FORSCOM Regulation 220-2, "Methods for Integrating Regular Army, Army National Guard, and Army Reserve Organizations," Headquarters, United States Army Forces Command (FORSCOM), Ft. Bragg, N.C., May 31, 2017.

[28] Memorandum dated December 1, 2015, subject: 101st Airborne MCU Headquarters Pilot Program Final Assessment Executive Summary; Memorandum dated December 16, 2015, subject: 101st Airborne MCU Assessment of Mission Effectiveness; AC senior field-grade officer, January 10, 2017, Ft. Campbell, Ky.; AC senior officer, telephone interview, January 11, 2017; Three AC officers, Ft. Campbell, Ky., January 9, 2017; Civilian in division mission support element, telephone interview, May 22, 2017. However, in many cases source materials and interviewees stated the operational gaps created by force reductions had been only partially closed.

[29] Robert B. Abrams, "FORSCOM Command Training Guidance (CTG) – Fiscal Year 2018," memorandum, Ft. Bragg, N.C., March 24, 2017.

cuts: AC division HQ are expected to be "ready now," which, in the worst case, could mean performing missions across the full spectrum of conflict without 20 percent of their authorized command post staffing during the first 270 days.

A specific example of a short-notice mission would be a GRF deployment. As described in a recent RAND Arroyo Center study examining scenarios that would require deployment of a brigade-sized GRF within a 96-hour time line,[30]

> the mandate for the GRF is contained in a Joint Chiefs of Staff executive order that codifies generalized global missions for which the GRF needs to be prepared, forces that could be called upon as part of the GRF (from across the Joint community), and time lines for providing them. The time lines, among other factors, make the GRF an important national asset for rapid responses to unforeseen or, more specifically, unplanned operations.[31]

Such deployments would require the kind of higher level mission command, possibly as a JTF, that is a function of a division HQ—although this does not necessarily mean the deployment of a full DTAC or MCP. Such scenarios might employ the type of home station command post envisioned by the FARG II designers as discussed above. Mobilizing for duty at the division HQ home station rather than deployment overseas could make a MCP-OD available sooner, depending upon assumptions about airflow and the speed of reception, staging, onward movement, and integration activities in theater.[32]

In response to a short- or no-notice emergency mission, the MCP-OD could be mobilized in 30 days. However, this would require specific SECDEF approval to waive notification policy. Furthermore, such a mobilization time line would mostly be "come as you are."[33] Any postmobilization training required would have to be completed before deployment.[34] Estimates by division HQ and MCP-OD leadership, as well as other subject-matter experts, of the number of full-time training days necessary to bring an MCP-OD soldier to deployment readiness ranged from 21 to 75 days.[35] Estimating the amount of training time needed is complicated both by variations in individual job requirements and training levels and by the way that MCP-OD readiness is reported.

Another potential challenge is the policy that limits the frequency of RC unit mobilizations at a slower rate than AC units.[36] The goal for AC units is to spend at least two years nondeployed between each year of deployment. In contrast, the goal for RC units is five nonmobilized years

---

[30] Christopher G. Pernin et al., *Enabling the Global Response Force Access Strategies for the 82nd Airborne Division*, Santa Monica, Calif.: RAND Corporation, RR-1161-A, 2017, p. 11.

[31] Pernin et al., 2017.

[32] In Chapter Five, we model several potential scenarios.

[33] It might be assumed that in the case of a contingency that required the rapid deployment of a full division, such a waiver would be forthcoming.

[34] Telephone interview with mobilization planner, July 19, 2017.

[35] Two AC MCP staff officers, Ft. Campbell, Ky., January 10, 2017; AC senior officer, telephone interview, January 11, 2017; National Guard senior field-grade officer, Ft. Bliss, Tex., February 12, 2017; AC division primary staff officer, Schofield Barracks, Hawaii, April 8, 2017; MCP-OD administrative officer, telephone interview, May 25, 2017; AC division primary staff officer, Ft. Bragg, N.C., June 13, 2017; AC HHBn primary staff officer, Ft. Bragg, Ky., June 14, 2017; AC operations section staff officer, Ft. Bragg, Ky., June 14, 2017.

[36] Six Combined Arms Center AC officers and Department of the Army Civilians, Ft. Leavenworth, Kan., January 4, 2017; Two AC MCP staff officers, Ft. Campbell, Ky., January 10, 2017.

to one mobilized year as the ideal for how frequently RC units are activated.[37] Although this ratio is a goal that can be waived by the Secretary of the Army to increase the RC rotation rate up to four to one, under current constraints an AC division could have its MCP-OD unavailable for at least every other deployment.[38] To avoid this constraint, a division commander could mobilize less than half of the MCP-OD as a derivative unit and avoid resetting the full unit's dwell requirement. (This might be a conscious choice or a collateral effect of tailoring their battle roster to fit the mission requirement, personnel caps, etc.)[39] As a practical matter, other MCP-ODs—including those from ARNG divisions—could be mobilized instead. However, this would obviate the value of premobilization collective training with the gaining division.

Readiness Reporting

As individuals, the personnel provided by the MCP-OD are an integral part of many sections and cells within a division's set of command posts. However, as an "AA" entity each MCP-OD submits its Commander's Unit Status Report (CUSR) independent of the division HQ with which it is partnered.[40] We observe two issues raised by this situation.

First, if the AC portion of the HQ is able to achieve the directed C-rating level (or the four measured areas of readiness—personnel, supply, maintenance, training—that comprise the C rating) without considering the MCP-OD, does such indicate that the MCP-OD is not a *requirement* for the division HQ to accomplish its full range of assigned missions?[41] Several interviewees suggested that if the MCP-ODs were counted on the AC division CUSRs, the divisions would be unable to report higher than T-2 or T-3 under the Objective T requirements currently being implemented, because it would not be possible to get a sufficient number of RC soldiers and leaders at key events.[42]

Similarly, how can a MCP-OD commander assess the unit's level of training (T rating) independent of the AC portion of division command posts it supports? The MCP-OD is not a unit in the sense of a similarly sized tank or infantry company because it does not conduct maneuver or other functions as a cohesive whole. As one field grade officer assigned to an MCP-OD dryly noted: "This is problematic because collective training requires integration with the MCP."[43] In practice, MCP-OD commanders, administrative officers,[44] and others stated that they reported

---

[37] See Robert Gates, Secretary of Defense, "Utilization of the Total Force," memorandum to the Secretaries of the Services, Chairman of the Joint Chiefs of Staff and the Undersecretaries of Defense, Washington D.C., January 19, 2007.

[38] Especially since most ARNG soldiers volunteer for duty in units known to be mobilizing, lifting this policy restriction could have merit. See the discussion in Chapter Three and recommendation in Chapter Five of Schnaubelt et al., *Sustaining the Army's Reserve Components as an Operational Force*, Santa Monica, Calif.: RAND Corporation, RR-1495-A, 2016.

[39] This kind of "economy of force" employment strategy was proposed to division staffs by individuals at FORSCOM, according to one staff officer, Ft. Campbell, Kentucky, January 10, 2017.

[40] Readiness reporting requirements are detailed in Headquarters, Department of the Army, *Army Pamphlet 220-1, Defense Readiness Reporting System—Army Procedures*, Washington, D.C., November 16, 2011.

[41] Unit status reports are classified; therefore, we do not address specific unit ratings (see Headquarters, Department of the Army, 2011, ch. 11). Nonetheless, the logic for questioning such a disparity should be obvious.

[42] Six Combined Arms Center AC officers and Department of the Army Civilians, Ft. Leavenworth, Kan., January 4, 2017. For an overview of Objective T standards, see Michelle Tan, "'Objective T': The Army's New Mission to Track Training," *Army Times*, October 11, 2016.

[43] MCP-OD officer, Ft. Bragg, N.C., June 14, 2017.

[44] When the commander is not active/guard reserve, the senior full-time officer performs similar functions from day to day as the unit administrative officer.

their T rating against the same staff METLs as did the AC portion of the HQ, but this did not fully resolve the assessment challenge.[45]

Organization

According to one unofficial definition of DOTMLPF-P, "Organization analysis examines how we are organize [sic] to fight; divisions, air wings, Marine-Air Ground Task Forces and other. It looks to see if there is a better organizational structure or capability that can be developed to solve a capability gap. Where is the problem occurring? What organizations is the problem occurring in? Is the organization properly staffed and funded to deal with the issue?"[46]

At its heart, the FARG II is an organizational change. It defined how many people, at what grades and with what skills, will make up a division HQ. As we described in Chapter Two, significantly fewer soldiers are now authorized in these HQ. FARG II also defined from which component each individual assigned to a division HQ will normally come.[47] The end state is not only fewer soldiers authorized in total but also fewer of the authorized soldiers coming from the AC.

Among the points made in our interviews and site visits were the following:

- All the liaison officer positions in the pre-FARG division HQ were moved over to the MCP-OD. A handful of interviewees suggested this could be a problem because they may not have the necessary familiarity with the division staff.[48]
- Specific MI MOSs may be harder to find or maintain in the RC. This may depend on the state that generates a specific ARNG MCP-OD and the distribution of MI slots within that state. Given limitations in the data available to us, we were unable to draw a clear conclusion regarding this concern. However, we also note that the AC may have a similar issue regarding the percentage of assigned compared to authorized MI positions in a division HQ.

Active Component/Reserve Component Integration

This subtopic could also be covered under doctrine and policy given that the Army's Total Force Policy directs the integration of "AC and RC forces and capabilities at the tactical level (division and below), consistent with the Secretary of Defense's policies for the use of the Total Force."[49] However, we address it here to focus on the issue arising from a specific organi-

---

[45] AC officers and Department of the Army Civilians, Ft. Leavenworth, Kan., January 4, 2017; ARNG field-grade officer, Ft. Bragg, N.C., June 14, 2017; Division intelligence section officer, Ft. Bragg, N.C., June 14, 2017.

[46] AcqNotes, "JCIDS Process: DOTMLPF-P Analysis."

[47] We use "normally" here because there is always the possibility that the right person will not be available from the "optimal" component, and the position will end up being filled by someone from another Army component. It can also be filled by personnel from another service or country, but at that point the headquarters will normally be serving as a combined or JTF.

[48] AC field-grade officer, Schofield Barracks, Hawaii, April 8, 2017; ARNG officer, telephone interview, May 25, 2017. For a description of liaison functions, see Headquarters, Department of the Army, FM 6-0, *Commander and Staff Organization and Operations*, May 5, 2014, ch. 13; Headquarters, Department of the Army, ATP 6-0.5, March 1, 2017, pp. 1-11–1-12. However, in the experience of one of the authors—who has served in several division HQ positions, including an assignment as division G-3 (operations officer)—liaison officers are typically chosen based upon whomever might be readily available at the moment.

[49] John M. McHugh, Secretary of the Army, Army Directive 2012-08 (Army Total Force Policy), September 4, 2012. Also, see Ellen M. Pint, Christopher M. Schnaubelt et al., *Review of Army Total Force Policy Implementation*, Santa Monica, Calif.: RAND Corporation, RR-1958-A, 2017.

zational change, not other relevant but less germane areas, such as total army training or career management.

Building a unit "culture" from this consolidated organization was frequently mentioned as an issue requiring a period of interaction between MCP-OD and AC division HQ personnel. More generally, the National Commission on the Future of the Army described the intercomponent tensions that exist and create cultural barriers within the Army.[50] However, culture is notoriously difficult to define and measure.

A recent RAND Arroyo Center review of research[51] conducted for the U.S. Army adopted the definition of culture provided by Ed Schein:

> Culture is a pattern of shared basic assumptions, invented, discovered, or developed by a given group as it learns to cope with its problems of external adaptation and internal integration that has worked well enough to be considered valid, and, therefore, is to be taught to new members of the group as the correct way to perceive, think, and feel in relation to those problems reframing.[52]

Many AC and ARNG personnel in our interviews mentioned organizational culture, and many described ways that they foster integration of MCP-OD soldiers into AC units. Common approaches used to foster integration include communication strategies and advanced planning, job assignments, and resourcing.

Several strategies mentioned in our interviews emphasized the need to communicate early and often to build relationships. Interviewees discussed a variety of efforts to connect with MCP-OD personnel six months or more before an exercise to get acquainted, assess talent, identify needs for training, share SOPs, complete administrative requirements (e.g., ensure that MCP-OD personnel have appropriate clearances), and otherwise prepare for the event. (The later "Training" section provides examples of these efforts and their effects on performance.) The language used to communicate also matters; for example, the 101st Airborne engaged with its ARNG units in a "partnership conference" in advance and stated that there is no component, because "everyone is part of one team."[53] Several interviewees in MCP-ODs reported that they were welcomed into the units; units were ready to receive them, and they are not treated differently from AC soldiers.[54]

Job preparation and assignments were another way of fostering cultural integration. The senior advisor to the adjutant general in the Texas ARNG used U.S. First Army–Division East personnel to prepare the MCP-OD for deployment; he was adamant that MCP-OD personnel who volunteered to deploy not be "bench players" and remain in garrison while the 1st Armored Division deployed.[55] Similarly, another officer stated that they did not routinely

---

[50] National Commission on the Future of the Army, *Report to the President and the Congress of the United States*, Arlington, Va.: U.S. Department of Defense, 2016.

[51] Meredith et al., *Identifying Promising Approaches to U.S. Army Institutional Change: A Review of the Literature on Organizational Culture and Climate*, Santa Monica, Calif.: RAND Corporation, RR-1588-A, 2016.

[52] Edgar H. Schein, "Organizational Culture," *American Psychologist*, Vol. 45, No. 2, 1990, pp. 109–119.

[53] AC senior officers, Ft. Campbell, Ky., January 10, 2017.

[54] ARNG field-grade officer, and others, Ft. Bragg, N.C., June 13, 2017.

[55] ARNG field-grade officer, Ft. Bliss, Tex., February 12, 2017.

assign ARNG personnel to night shift only[56] or hesitate to send out ARNG personnel for high-profile tasks. An ARNG officer supporting the 82nd Airborne Division (82nd Airborne) said they had a key role in the WFX and that their AC counterpart asked for and appreciated their help.[57] In contrast, another interviewee who did not feel well integrated reported that he was told that he was not needed because the MCP was overstrength, and he was instead assigned to work in the MEB in the SACP.

A third strategy consisted of providing MCP-OD personnel with resources, both tactical and symbolic. For example, one MCP-OD officer reported that the HQ battalion provided resources and treated the MCP-OD as an AC unit. This included incorporating a parachute jump in training for MCP-OD personnel and providing them with patches and berets.[58] An ARNG officer stated, "We wear the patch. No one knows I'm Guard until I tell them. I feel like a part of the team."[59] An NCO from a MCP-OD emphasized the need to provide the right equipment and uniforms to make personnel feel like part of the group; in this case, it took some time to obtain appropriate resources, and some MCP-OD personnel bought uniforms[60] and berets with their own money. One interviewee, however, reported that giving patches to the MCP-OD goes against FORSCOM policy (although the division issued patches regardless). The respondent questioned why FORSCOM would have a policy making it more difficult to integrate these personnel.[61]

Several interviewees from MCP-ODs said that they felt like they were "part of the team,"[62] and interviewees from both the AC and MCP-ODs reported that, once integrated, ARNG and AC soldiers were indistinguishable and interchangeable.[63] However, other interviewees stated that having different chains of command in the AC and RC causes friction, so it is important to find commonalities to support integration.[64] In only one case did an AC leader specifically say that having nobody would be better than having an RC MCP-OD,[65] but several other interview participants revealed negative attitudes toward the MCP-OD concept.[66]

An in-depth consideration of AC-RC cultural differences (or ARNG-USAR differences) falls beyond the scope of this study. However, we note that component and unit culture is widely perceived as a potential issue, and leaders from all components should recognize the need to build teams across components. To conclude with one example, the RAND Arroyo Center study team interviewed 31 soldiers in the 82nd Airborne and its MCP-OD. "Should

---

[56] The night shift is often presumed to have a slower pace than the day shift.

[57] AC field-grade officer, Ft. Bragg, N.C., June 13, 2017.

[58] ARNG field-grade officer, Ft. Bragg, N.C., June 13, 2017.

[59] ARNG field-grade officer, Ft. Bragg, N.C., June 13, 2017.

[60] ARNG field-grade officer, Ft. Bragg, N.C., June 13, 2017.

[61] AC field-grade officer, Ft. Bliss, Tex., January 10, 2017.

[62] ARNG field-grade officers and others, Ft. Bliss, Tex., February 13, 2017; Ft. Bragg, N.C.

[63] AC field-grade officer and NCO, Ft. Bragg, N.C., June 14, 2017.

[64] AC field-grade officer, Ft. Bliss, Tex., February 14, 2017.

[65] AC field-grade officer, June 14, 2017, Ft. Bragg, N.C.

[66] For example, AC company-grade officer, Ft. Campbell, Ky., January 9, 2017; AC field-grade officer, February 15, 2017, Ft. Bliss, Tex.; AC division primary staff officer, Ft. Bragg, N.C., June 13, 2017; AC NCO, Ft. Bragg, N.C., June 13, 2017; AC field-grade officer, June 14, 2017, Ft. Bragg, N.C.

MCP-OD personnel be airborne qualified?" was part of our standard question set for both sets of soldiers. In every case in which the interviewee offered an opinion, except one, the answer was yes, because being jump qualified was necessary to fit into the unit culture.[67]

U.S. Army Reserve Fills

One of the unusual features of the MCP-OD MTOEs is that they have documented requirements for CA and MISO (formerly called Psychological Operations) soldiers, but they do not have the accompanying authorizations. This is presumably because the ARNG lacks the unit structure to generate significant numbers of CA or MISO soldiers, and therefore any MCP-OD positions in these career fields would be hard to fill and sustain. The expectation is that these requirements would be filled by the USAR, but only when the MCP-OD is notified of a pending mobilization. To date, these positions have largely been filled by personnel from within the U.S. Army Civil Affairs and Psychological Operations Command HQ, to avoid breaking up subordinate units.[68]

Ironically, this ad hoc sourcing solution weakens one of the likely advantages of the MCP-OD concept over other options (such as a more modular system). If CA and MISO soldiers are only identified after the MCP-OD is notified of its pending mobilization, they gain none of the experience working with the AC HQ in weekend and longer training sessions. For example, at the 82nd Airborne's WFX, the additional CA positions in the command post were filled by AC CA soldiers, conveniently located at Fort Bragg.[69]

If MCP-ODs remain partner units, instead of becoming part of a consolidated unit structure, there are several ways the current sourcing solution might be modified to ensure a more ready USAR contribution to the division HQ:

- The MCP-OD MTOE could be converted to a true MCU and USAR soldiers assigned to fill the CA and MISO positions. This might have the secondary benefit of providing training opportunities for CA and MISO soldiers who happen to move away from USAR units into areas closer to the MCP-OD location.[70]
- MTOEs for USAR CA battalions and MISO companies, which by doctrine each support a division, could be modified to have one or more dedicated "MCP-OD cell" positions that could have training events more closely aligned with the division than the rest of the battalion. The advantage would be that these positions could more easily be filled from within the unit; the disadvantage is that the difference in training and mobilization plans would make this an awkward "unit within a unit."
- The Army could formalize the use of U.S. Army Civil Affairs and Psychological Operations Command (USACAPOC) and create a MCP-OD support structure, authorizing all assigned soldiers to perform "duty at" the division HQ. Because of the high percentage of full-time personnel at U.S. Army Civil Affairs and Psychological Operations Com-

---

[67] AC senior officer, Ft. Campbell, Ky., January 10, 2017; Multiple interviews, Ft. Bragg, N.C., June 13–16, 2017.

[68] Telephone interview with Army Reserve personnel involved in sourcing and mobilization, August 7, 2017.

[69] This is also an example of our earlier observation that division command posts tend to be augmented beyond MTOE positions for WFX.

[70] This could raise new issues of having Title 10 USAR soldiers reporting to a Title 32 ARNG MCP-OD commander prior to mobilization. Alternatively, the USAR soldiers might report independently to the division's AC chief of staff.

mand (USACAPOC) and the small number of MCP-OD personnel, this should not impose a significant management burden on the HQ but would allow trained personnel to be rotated into the positions and train with the division HQ well before mobilization notification.

- Army Reserve Elements or similar organizations providing USAR support to installations or AC commands might also provide MTOE or TDA structures for authorizing CA and MISO positions.
- Require the ARNG to train and assign CA and MISO soldiers to the MCP-OD positions. However, it could be a major challenge to recruit, train, and sustain soldiers for these positions absent a larger force-management decision to create ARNG CA and MISO units from which the MCP-OD could draw.

Training

According to acquisitions experts,

> training analysis examines how we prepare our forces to fight tactically from basic training, advanced individual training, various types of unit training, joint exercises, and other ways to see if improvement can be made to offset capability gaps. Is the issue caused, at least in part, by a complete lack of or inadequate training? Does training exist that addresses the issue? . . . Leadership and education analysis examines how we prepare our leaders to lead the fight from squad leader to 4-star general/admiral and their overall professional development.[71]

Once personnel have been assigned to the HQ and while they are receiving continued human resources (HR) support, the Army must train and develop them as individuals and teams to conduct their peacetime and wartime functions. Most of our interview subjects highlighted this as the most critical of DOTMLPF-P functions for the new structure. In this section, we assess the impact of the FARG II design on initial individual skill training and continued development and collective training.

Initial Individual Skill Training and Continued Development

As will be discussed in the "Personnel" section below, most AC personnel will have completed their MOS-qualifying training before being assigned to a position in a division HQ. RC soldiers, however, are typically assigned to a unit while attending such training. It is therefore common for RC units to have lower DMOSQ rates than AC units. Many interviewees commented that MCP-OD personnel are not MOS qualified;[72] although 100 percent (or more) of "bodies" were assigned to the unit, the unit was operating at a much lower effective strength.[73] The lack of DMOSQ personnel was especially problematic for MI positions and in low-density MOSs, such as public affairs, simulations, and information operations.[74]

---

[71] AcqNotes, "JCIDS Process: DOTMLPF-P Analysis."

[72] ARNG and AC field-grade officers, Ft. Bragg, N.C., June 13–14, 2017; AC NCO, Ft. Campbell, Ky., January 10, 2017

[73] ARNG field-grade officer, Ft. Bragg, N.C., June 13, 2017

[74] ARNG field-grade officer, Ft. Bragg, N.C., June 14, 2017.

It is important to also highlight that these DMOSQ gaps are not consistent across states within either RC. Some states have many MI spaces, which have created over time a pool of MI soldiers with varied MOSs and experiences; a MCP-OD based in these states will tend to have less trouble finding personnel with the right MOSs and clearances than an identical unit in a state with limited MI positions.

Even when MCP-OD soldiers are MOS qualified, several interviewees reported that the soldiers lack experience. Personnel need repeated exposure to their peacetime and wartime tasks to be considered ready at any point in time, and some interviewees[75] raised concerns about the lack of opportunities for USAR and ARNG personnel to practice perishable skills: "If you don't use it, it goes away." Opportunities to conduct key tasks, however, can be limited, even in the AC. For example, a recent RAND study of all-source analysts (35F MOS) in the AC found that both new analysts (new Advanced Individual Training graduates) and midgrade analysts (E6s) had few opportunities to perform key job tasks, particularly in garrison (as compared to unit field training and deployment settings).[76] A related point discussed in the section on materiel is that MCP-OD personnel often do not have opportunities to train on the systems used during the WFX. Some responses indicated that personnel could be up to speed quickly (e.g., within a couple of days), but this could take longer, depending on the type of equipment used on the job.[77]

Use of Command Post of the Future (CPOF) equipment was frequently mentioned as a specific RC training shortfall.[78] Interviewees, from both AC and RC, stated that this was a recurring issue that hindered quick RC soldier integration into division command posts. Interviews described the CPOF and other IT systems, such as Distributed Common Ground System–Army (DCGS-A), as having a steep learning curve, with one interviewee reporting that it takes 14 days to train AC soldiers on the CPOF Agile Client.[79] A related point is that technology changes rapidly, but drill time does not allow the RC to update their skills.[80] Respondents pointed to the need for ongoing training on CPOF, such as integrating CPOF training into RC units' yearly training plans;[81] this training could be supported by mobile training teams and other training resources, but it would also require that MCP-ODs have the equipment needed to conduct this training.[82]

---

[75] ARNG field-grade officer, Ft. Bliss, Tex., February 12, 2017; AC company-grade officer, Ft. Bliss, Tex., February 14, 2017.

[76] Lytell et al., *Assessing Competencies and Proficiency of Army Intelligence Analysts Across the Career Life Cycle*, Santa Monica, Calif.: RAND Corporation, RR-1851-A, 2017.

[77] AC NCO, Ft. Campbell, Ky., January 10, 2017; AC warrant officer, Ft. Bragg, N.C., June 16, 2017.

[78] CPOF is the Army's primary system for viewing and sharing mission command information.

[79] ARNG field-grade officers, Ft. Bliss, Tex., February 12–13, 2017; AC field-grade officer, Ft. Bliss, Tex., January 10, 2017; ARNG NCO, Ft. Bragg, N.C., June 14, 2017; AC field-grade officer, Ft. Campbell, Ky., January 10, 2017; AC NCO, Ft. Bliss, Tex., February 13, 2017; AC division primary staff officer, Schofield Barracks, Hawaii, April 8, 2017; ARNG NCO, Ft. Bragg, N.C., June 13, 2017. DCGS-A "Provides distributed intelligence, surveillance and reconnaissance (ISR) planning, management, control and tasking; multi-intelligence fusion." See U.S. Army Acquisition Support Center, "Distributed Common Ground System–Army."

[80] AC NCO, Ft. Bliss, Tex., February 13, 2017.

[81] AC division primary staff officer, Schofield Barracks, Hawaii, April 8, 2017; ARNG field-grade officer, Ft. Bliss, Tex., February 13, 2017.

[82] AC field-grade officer, Ft. Bliss, Tex., January 10, 2017; ARNG field-grade officer, Ft. Bragg, N.C., June 13, 2017; ARNG field-grade officer, interview conducted by telephone on May 25, 2017.

The Mission Command Training Program and Preparing for the Division Warfighter

The U.S. Army CAC's Mission Command Training Program (MCTP), which orchestrates the execution of the aforementioned WFX, has emerged in past decades as the premier combat training center to train brigades, divisions, corps, and Army service component command (ASCC)–level HQ on their mission essential tasks needed to support the unified land operations.[83] The MCTP origins lie with the establishment of the Battle Command Training Program (BCTP) in 1987. By design, its original function was to improve "battlefield command and control through stressful and realistic combined arms training in a [computer simulated] combat environment."[84] With the end of the Cold War, as well as in analyzing lessons learned from the Gulf War, the U.S. Army deliberately sought to broaden the BCTP's mission sets to prepare the military to address the world's emerging military issues and challenges that were linked to the rapidly transforming global-political environment. Much effort and resources were then allocated to expand both the curriculum and the number of personnel and systems needed to support such an expansion. On May 10, 2011, the BCTP was redesignated as the U.S. Army MCTP.

From its conception as the BCTP to the time of publication, the MCTP has always featured elements and characteristics of the combat training center training model, such as "free-thinking" opposing force, the use of observer/trainers, and the oversight of the exercise's organized sequence of events from the start of exercise to the culminating after-action review (AAR).[85] Unique to the MCTP is its heavy reliance on computer simulation, mobile observer trainer teams, and the assignment of senior mentors (all retired Army general officers) for unit commanders. In any given year, the MCTP supports on average: 5x Multi-Echelon, Multi-Component Warfighter Exercises; 5x Army Service Component Command Exercises; and 6x National Guard Brigade Warfighters.[86]

It can very well be argued, therefore, that to a division-commanding general there is no greater test outside combat of a division HQ than that of a MCTP WFX. In preparation for the warfighter, the division HQ allocates time to conduct prewarfighter training events (e.g., one or more computer-simulation command post exercises [CPXs]). The division HQ ensures that each CPX (at its core) will exercise all facets of the division staff and provides the necessary stressors that will help temper the staff's ability to effectively "fight" during the upcoming warfighter. These CPXs, more so than the associated "sergeant's time" training and section-level training events, are critical events that enable the successful laying of a strong and level foundation upon which the "team" is built. The CPX's importance lies not only in ensuring that each member knows their respective duties but also in ensuring that each member is familiar with the strengths and weakness of their fellow members and surrounding sections.

It is in this light that a capability gap becomes apparent with respect to the RC soldiers supporting the MCP-OD. This gap is due in part to the limited time of AT, which effectively constricts the RC soldier to take part only in either the WFX or CPX (let alone multiple CPXs). This concerning gap and the issues stemming from it are discussed further in later sections.

---

[83] Mission Command Training Program, U.S. Army Combined Arms Center, "Frequently Asked Questions."

[84] Mission Command Training Program, U.S. Army Combined Arms Center, "Mission Command Training Program History."

[85] Mission Command Training Program, U.S. Army Combined Arms Center, "Mission Command Training Program History."

[86] The Mission Command Training Program, U.S. Army Combined Arms Center, "MCTP Orientation Brief."

Collective Training

Even a set of well-trained individuals may not work well as a team until they have trained together. In this category, we include both sectional training, as when the G2 trains a small group of soldiers to produce together a complex intelligence product, and multisectional training, as when the full MCP goes to the field and operates as a unit.

As described in the section on organization, training together is a factor that contributes to integrating AC and RC units. Interviewees also mentioned other ways in which training together are important for effectiveness. One officer reported that MCP-OD troops are as capable as the average AC soldier, but the major shortfall is the time available to train on collective tasks and to integrate.[87] Understanding SOPs, commander's intent, strengths and weakness of personnel, and the way "things work" in a particular command post requires time for MCP-OD and AC division HQ personnel to interact and learn. As an operations officer described it, the most important aspect in the MCP is developing an "understanding of the CG's approach to operations."[88] Interestingly, one ARNG officer reported that it is more difficult to integrate "the more senior you are. In the middle of a warfighter exercise, you don't know the key players and don't know the battle rhythm."[89]

Several subjects gave specific examples of the value of understanding others' strengths and weaknesses. While these are hard to describe concisely in this context, the theme was that knowing the other people in the HQ is crucial to problem solving and staff success. Illustrating what psychologists term a "transactive memory system," teams develop shared mental models about "who knows what." Such collective knowledge is significantly and positively related to team performance, learning, and creativity, especially when teams work on nonroutine tasks or in turbulent settings.[90] In the case of the MCP-ODs, this emphasizes the importance of training team members together on their tasks rather than individually, taking time to plan work to identify each team member's expertise, and composing stable teams.[91] Thus, although the Army is built largely on a concept of interchangeable parts, teams perform more effectively when members have opportunities to learn about others' knowledge and skills. One AC officer in our interviews stated, "We need to know our soldiers—how to

[87] AC field-grade officer, Ft. Bragg, N.C., June 14, 2017.

[88] AC division primary staff officer, Schofield Barracks, Hawaii, April 8, 2017.

[89] ARNG field-grade officer, Ft. Bliss, Tex., February 12, 2017.

[90] See Ali E. Akgün et al., "Knowledge Networks in New Product Development Projects: A Transactive Memory Perspective," *Information and Management*, Vol. 42, No. 8, 2005, pp. 1105–1120; John R. Austin, "Transactive Memory in Organizational Groups: The Effects of Content, Consensus, Specialization, and Accuracy on Group Performance," *Journal of Applied Psychology*, Vol. 88, No. 5, 2003, p. 866; Gino et al., "First, Get Your Feet Wet: The Effects of Learning from Direct and Indirect Experience on Team Creativity," *Organizational Behavior and Human Decision Processes*, Vol. 111, No. 2, 2010, pp. 102–115; Diane Wei Liang, Richard Moreland, and Linda Argote, "Group Versus Individual Training and Group Performance: The Mediating Role of Transactive Memory," *Personality and Social Psychology Bulletin*, Vol. 21, No. 4, 1995, pp. 384–393; Yuqing Ren, Kathleen M. Carley, and Linda Argote, "The Contingent Effects of Transactive Memory: When Is It More Beneficial to Know What Others Know?" *Management Science*, Vol. 52, No. 5, 2006, pp. 671–682.

[91] See Liang, Moreland, and Argote, 1995, 384–393; Glenn Littlepage, William Robison, and Kelly Reddington, "Effects of Task Experience and Group Experience on Group Performance, Member Ability, and Recognition of Expertise," *Organizational Behavior and Human Decision Processes*, Vol. 69, No. 2, 1997, 133–147; Diane Liang Rulke and Devaki Rau, "Investigating the Encoding Process of Transactive Memory Development in Group Training," *Group and Organization Management*, Vol. 25, No. 4, 2000, 373–396.

motivate them, how people learn better. We know our soldiers [but do not know the ARNG soldiers as well]."[92]

One officer explicitly connected this to the questions of nonaligned dwell-time policies, saying,

> After I put in all of the work to bring someone on board and get them fully integrated, I can use them once, but then they are off limits for months. If they are deployed for one year, they then can't be deployed again for 40 months. After 12 to 24 months, all of the RA [Regular Army] people they worked with and integrated into have moved on. MCP-OD personnel have to relearn tasks and people. They'd be OK with relearning the tasks, but relearning the people would be harder.[93]

Comments from many interviewees indicated that opportunities to interact prior to the WFX was an important factor in collective training (or conversely, that a lack of opportunities to interact impeded learning to work as a team). For example, one interviewee reported that by showing up only three days before the WFX, ARNG soldiers had to "learn the basics," which interfered with learning to work together. Two[94] interview participants commented on the need to be involved in military decisionmaking process prior to the exercise. In contrast, another participant described benefits of MCP-OD personnel arriving at the WFX a week in advance to get up to speed on MI systems.[95] Others described the value of AC and MCP-OD engagement further in advance of field training exercises to touch base, receive orders and missions, share SOPs, develop a shared understanding of "jargon," understand roles and expectations, determine how to incorporate MCP-OD personnel into sections, and identify the skills and resources (e.g., clearances) that MCP-OD personnel would need to help them get up to speed and "into the fight."[96] Individuals who participated in the 101st Airborne's multicomponent HQ deployment in 2016 highlighted the weekly calls between the division and the state ARNG HQ to review training statistics[97] and that the division budgeted approximately $250,000 for travel by the ARNG members to the division's installation, or vice versa, before deployment.[98] A theme underlying these preliminary activities was the need for advance planning on the part of the AC and RC.

Finally, both MCP-OD and AC leaders played a key role in training and integration. As described in the section on organization, senior leaders set expectations about integrating MCP-OD personnel (e.g., via job assignments). Several interviewees credited officers, NCOs, and warrant officers with integration strategies, such as advance planning and communication, as well as serving as mentors to MCP-OD personnel and helping them work through technical

---

[92] AC NCO, Ft. Bliss, Tex., February 13, 2017.

[93] AC field-grade officer, Ft. Bragg, N.C., June 13, 2017.

[94] ARNG field-grade officer, Ft. Bliss, Tex., February 12, 2017; AC field-grade officer, Schofield Barracks, Hawaii, April 10, 2017.

[95] AC NCO, Ft. Bliss, Tex., February 13, 2017

[96] AC company-grade officer and NCO, Ft. Bliss, Tex., February 14, 2017; AC field-grade officer, Ft. Bragg, N.C., April 10, 2017; AC field-grade officer, Ft. Bliss, Tex., February 13, 2017; ARNG field-grade officer, Ft. Bliss, Tex., February 13, 2017.

[97] AC warrant officer, Ft. Campbell, Ky., January 10, 2017.

[98] AC field-grade officer, Ft. Campbell, Ky., January 10, 2017. Unlike most MCP-ODs, the units in the MCU pilot program were not the in same state, increasing the cost of this valuable interaction.

and administrative challenges, although some respondents identified the need for more warrant officers in MCP-ODs to coordinate with division counterparts. One AC officer singled out an ARNG colonel for playing a critical role in advising and mentoring ARNG soldiers personnel across the staff and advising the command on placement of individuals to maximize the fit of their talents with the organization and mission.[99]

Evaluating a Division Headquarters

The objective of training is obviously to prepare the HQ (or any unit) to be utilized in some way—a "real-world mission." Progress in this process of preparation is measured in "readiness." How to measure unit readiness is a longstanding challenge to the Army and goes far beyond the scope of this paper. Some understanding of what is meant by readiness is crucial, however, because many would say readiness is the ultimate metric for assessing the success of the FARG II design.

The Army (and the other armed services) reports the readiness of units based upon category, or C-level, ratings. As the Chairman of the Joint Chiefs of Staff (CJCS) *Guide to the Chairman's Readiness System* explains, "The C-level reflects the status of the selected unit resources measured against the resources required to undertake the wartime missions for which the unit is organized or designed."[100] These are summarized in Table 4.1.

The conundrum facing the Army is how to apply these to the current HQ as an organization. If "the unit" is defined as the RA HQ MTOE, it can never be above C-2, because the MCP-OD is clearly a "compensation for deficiencies" in its capacity. If "the unit" is the fully formed HQ, evaluating its readiness (including training, manning, and DMOSQ) would require accessing the RC personnel assigned and having them present when training is evaluated. The Army has chosen the former but at the time of our interviews had not established how to overcome the logical inconsistency in that choice.

Synchronization of Active Component and Reserve Component Training Management Cycles

Synchronization of AC and RC training management cycles is a significant challenge in providing opportunities for collective training.[101] Synchronization of calendars and the amount of time available for training in the RC were among the issues that interviewees raised most frequently with respect to training.[102]

Whereas AC units perform long-range planning, especially for major exercises, they tend to plan largely on a *quarterly* basis. An exercise might be planned for a tentative range of dates a year in advance, but the dates might shift 90 days out. In contrast, RC units tend to plan training on a *yearly* basis. Among other reasons, this time frame allows maximum notice to civilian employers who are affected by the temporary absence of their employees who serve in the RC. When training dates are changed without several months of notice, RC units might not be able to participate.

---

[99] AC field-grade officer, Ft. Campbell, Ky., January 10, 2017.

[100] Joint Chiefs of Staff, *CJCS Guide to the Chairman's Readiness System*, CJCS Guide 3401D, Washington, D.C., November 15, 2010, p. 9.

[101] Further details on RC training requirements can be found in the appendix.

[102] Appendix A provides an explanation of RC duty codes and compares AC and RC training management cycles.

Table 4.1
Explanation of C-Level Ratings

C Level	Interpretation
C-1	The unit possesses the required resources and is trained to undertake the mission for which it is designed (that is, accomplish core functions and provide designed capabilities). The status of resources and training will neither limit flexibility in methods for mission accomplishment nor increase vulnerability of unit personnel and equipment. The unit does not require any compensation for deficiencies.
C-2	The unit possesses the required resources and is trained to undertake most portions of the mission for which it is designed (that is, accomplish core functions and provide designed capabilities). The status of resources and training may cause isolated decreases in flexibility in methods for mission accomplishment but will not increase the vulnerability of the unit under most envisioned operational scenarios. The unit would require little, if any, compensation for deficiencies.
C-3	The unit possesses the required resources and is trained to undertake many, but not all, portions of the mission for which it is designed (that is, accomplish core functions and provide designed capabilities). The status of resources or training will result in a significant decrease in flexibility for mission accomplishment and will increase the vulnerability of the unit under many, but not all, envisioned operational scenarios. The unit will require significant compensation for deficiencies.
C-4	The unit requires additional resources or training to undertake its designed mission (that is, accomplish core functions and provide designed capabilities), but it may be directed to undertake some portions of its mission with resources on hand.
C-5	The unit is undergoing a HQDA-directed resource action (for example, reconstitution) and is not prepared, at this time, to undertake the full-spectrum mission for which it is designed (that is, accomplish core functions and provide designed capabilities). However, it may be capable of undertaking nontraditional or nonstandard missions.

SOURCE: Headquarters, Department of the Army, November 16, 2011, para. 3-5.

Even when dates do not change, RC units will require a longer time span to prepare. As one interviewee noted, "[our division] published the WFX operations order 45 days before the exercise, but training time before the WFX was not built into the RC schedules."[103] If the operations order had instead been published six months in advance, RC personnel would have had several IDT periods to participate in the military decisionmaking process and work on staff estimates prior to the WFX.

RC units typically schedule their IDT on weekends to minimize the impact on civilian employers. Except for exercises and special events, AC units typically do not schedule training on weekends to limit operations tempo and minimize the impact on families. However, several ARNG soldiers interviewed argued for modifying these practices to increase the opportunities for essential cross-component training, such as having the MCP-OD drill during the week or having AC division sections work a weekend every other month (when the MCP-OD was training).[104]

Interviewees mentioned other factors contributing to aligning training calendars with respect to MI personnel. The schoolhouse has a limited number of accredited sites to conduct 35-series training, making it difficult to align with ARNG training days, and institutional training at Fort Huachuca is three months long, which conflicts with the timing of

---

[103] AC division primary staff officer, Schofield Barracks, HI, April 8, 2017.

[104] For example, ARNG field-grade officer, Ft. Bragg, N.C., June 14, 2017.

the WFX.[105] In addition, G2 personnel[106] reported that the total amount of training time for ARNG soldiers is insufficient; for example, the training required to learn DCGS-A exceeds available training days, and signals intelligence soldiers need to participate in two-week, online training courses annually to be current with National Security Agency standards. More generally, many of the personnel we interviewed reported that 39 training days was not sufficient for the RC to meet training requirements.

Materiel

"The materiel analysis examines all the necessary equipment and systems that are needed by our forces to fight and operate effectively and if new systems are needed to fill a capability gap. Is the issue caused, at least in part, by inadequate systems or equipment?"[107]

The study team expected this to be one of the least challenging areas of the transition to the FARG II design, because it did not directly affect the equipment assigned to the division HQ. It is relevant, however, because the existing practices for materiel acquisition and distribution may need to be modified to enable the HQ to be mission ready as quickly as possible.

For this report, we focus on how equipment is allocated to the different parts of the HQ and particularly the materiel required by the RC to train and sustain their readiness. For example, if ARNG intelligence soldiers would ideally be training on a particular system every month at their home station, but that system is not fielded to ARNG units, then what initially might seem a training shortfall is actually an equipping issue. We did not look at whether specific systems are most appropriate for a division HQ overall.

Access to information systems was the primary material challenge reported by interviewees. There were two parts to this issue: (1) computers and (2) networking. These parts are obviously related, but potential solutions may be different.

Access to Computers

Two interviewees mentioned a shortage in computer equipment for training at their ARNG home station.[108] This is particularly critical because MCP-OD personnel may be proficient with older systems, such as high-capacity line-of-sight radios but have not trained on more modern systems fielded in the current AC division HQ, such as the Warfighter Information Network-Tactical Increment 2 (WIN-T Inc 2).[109]

Access to Networks

A more common and apparently unresolved issue was access to networks, and some interviewees reported issues regarding the ability of RC soldiers to access AC networks and systems. Problems ranged from a lack of necessary clearances to obtain accounts for classified networks, to lack of training on the use of various systems, to loss of account access due to infrequent

---

105 AC G2 officer; Ft. Campbell, Ky., January 9, 2017.

106 AC G2 officer; Ft. Campbell, Ky., January 9, 2017.

107 AcqNotes, "JCIDS Process: DOTMLPF-P Analysis."

108 ARNG field-grade officer, Ft. Bragg, N.C., June 14, 2017.

109 AC NCO, Ft. Campbell, Ky., January 10, 2017.

log-ins. The latter was the most commonly reported problem. It apparently stems from networks routinely denying access if a user has not logged in during the previous 30 days. Restoring access requires action by a network administrator.[110] Many MCP-ODs schedule their training assemblies every other month to train for 4 days in a row rather than only the 2 days of a weekend. This technique may increase training value and exposure to AC counterparts but also results in soldiers losing their network access between training periods.[111]

Leader Development

> The leadership and education analysis examines how we prepare our leaders to lead the fight from squad leader to 4-star general/admiral and their overall professional development. Does leadership understand the scope of the problem? Does leadership have resources at its disposal to correct the issue?[112]

Among the hardest areas to assess is the role of leader development in the fielding of the MCP-ODs. On one hand, they are being created after 15 years of intense cross-component interaction, when most senior officers and NCOs have worked with soldiers from the other components. On the other hand, to the degree operational interactions decline over the next decade, the MCP-OD system will give subsequent cohorts regular, structured opportunities to interact with the other components. One example of such positive effects can be seen in the words of an AC captain who had to fill a lieutenant colonel position and stated,

> I've learned from the RC LTC [he worked with] a lot more about leadership—[for example,] interacting with higher echelon people and the different ways you address [them]—than in any other place. It has been a wonderful experience, not a detriment. It helps integration of the citizen soldier and Army grunt.[113]

There will be challenges to leader development along this path, however. Among those suggested or implied in our interviews would be the following:

- For AC leaders:
  - integrating subordinates from the RCs into established AC teams
  - training AC subordinates to quickly integrate RC personnel into their teams and routines
  - understanding the RC training management cycle, how it differs from the AC cycle, and planning accordingly for MCP-OD inclusion in training events
  - understanding the legal and policy requirements for mobilization and deployment of RC soldiers and how they affect MCP-OD availability without sufficient advance notification and planning.

---

[110] ARNG field-grade officer, Ft. Bragg, N.C., June 14, 2017.

[111] AC soldiers have also reported issues with connection to portals on installations other than their home station (AC NCO, Ft. Bliss, Tex., February 14, 2017).

[112] AcqNotes, "JCIDS Process: DOTMLPF-P Analysis."

[113] AC company-grade officer, Ft. Bliss, Tex., February 14, 2017.

- For RC leaders:
  - ensuring readiness of assigned soldiers when the optimal training facilities and events are located at some distance from their homes and/or RC center
  - identifying methods and opportunities to integrate MCP-OD training with AC training management cycles.

Personnel

> The personnel analysis examines availability of qualified people for peacetime, wartime, and various contingency operations to support a capability gap by restructuring. Is the issue caused, at least in part, by inability or decreased ability to place qualified and trained personnel in the correct occupational specialties? Are the right personnel in the right positions (skill set match)?[114]

While the personnel function can include a host of HR management activities, in this report we are most concerned about the process of assigning personnel to positions in the division HQ from any of the three components. A secondary concern is how these assignment actions are seen to fit into the broader process of career management, including how long each component sees the assignment lasting and how it contributes to the individual's retention, promotion, and development. Finally, there are concerns about how HR actions are accomplished in a multicomponent partnership.

It is important to note again the differences between AC and RC manning systems. The AC is a push system, where a centralized HR command orders soldiers from its national inventory to individual training and then to specific installations, where they are then allocated to units based on requirements. The RC is a pull system, where recruiters and units seek to attract personnel to fill their vacancies, and they often sign in unqualified for the position they fill. For this reason, the question of assigned compared to DMOSQ soldiers is a much larger issue for the MCP-OD than for its supported AC HQ.

Although the scope of this project focuses on the MCP-ODs, we note that MOS fill is not just an RC issue; it is also an AC issue. If a skill set in chronic short supply in the AC is readily available from an RC, their different demographics may complement each other and give the combined unit the necessary skills it needs for many contingencies. If, on the other hand, the different components have the same personnel challenges, the problems are compounded. Our analysis, presented first for the AC and then the ARNG, shows no consistent pattern. MI positions seem to be hardest to fill in both components; some MOSs such as the 42 series are shortages for the ARNG but not the AC; and some are the reverse (91C utilities equipment repairer would seem to be a classic MOS that would draw from RC-heavy civilian skills). CA, filled in the MCP-ODs from the USAR, presents a similar case where an AC shortage should be mitigated by fills from the nationwide USAR CA command.

Division Headquarters Fill

Unlike most of the other DOTMLPF-P areas, personnel challenges can be studied quantitatively, even as the division HQ and MCP-ODs are adjusting to the new design. We measured personnel fill as the ratio of assigned to authorized personnel by MOS/AOC (Army Operat-

---

[114] AcqNotes, "JCIDS Process: DOTMLPF-P Analysis."

ing Concept) and grade. For the analysis of fill rates at division HQ, we obtained FMSWeb MTOE authorizations ("spaces") from FY 2008 to FY 2016 for multiple division HQ. We also obtained Total Army Personnel Database (TAPDB) data ("faces") that included the duty military occupational specialty, individual's primary MOS/AOC, and individual's grade (PG) of enlisted personnel, warrant officers, and officers in those division HQ from FY 2008 to FY 2016.

We focused on identifying those MOSs/AOCs that had, over time (generally 2008–2016), average fill rates below 90 percent (an indicator of a personnel readiness challenge)[115] and that were on recent (2013 and later) MTOE documents for the division HQ. In particular, we focused on MOSs/AOCs—other than branch immaterial positions (e.g., MOSs 01A, 01B, 02A)—that were (a) on a division HQ MTOE for at least three recent years between FY 2008 and FY 2016; (b) had an average fill less than 90 percent (the minimum fill rate for a P-1 readiness level) across fiscal years; and (c) had an average fill less than 90 percent in the most recent fiscal year with data. We also looked at how much those fill rates tended to fluctuate within recent years and whether they improved during a deployment.

Appendix B contains tables summarizing fill rate patterns in division HQ over time. We identified MOSs/AOCs with personnel fill challenges that include average fill rates below 90 percent over time and fluctuating fill rates during recent fiscal years. (To identify fluctuating fill rates during fiscal years, we looked at the coefficient of variation—standard deviation of monthly fill rates for an MOS/AOC divided by the average fill rate for the MOS/AOC.)

These figures show that the Army seems to have habitually been willing to accept risk in manning of division HQ before deployment. We also see that, in general, the Fires and Intel cells are areas where the Army is less likely to take risk with AC manning. Furthermore, most issues seem to be in low-density MOSs. Regardless of the presence or absence of a ready or not-ready MCP-OD, Army divisions should expect to be alerted and potentially deploy with gaps in personnel and skills (capability and capacity). The systems that are in place to mitigate or address these challenges may also be of use in mitigating/addressing MCP-OD shortages/delay.

Main Command Post–Operational Detachments Fill

When analyzing MCP-OD data, we did not look at fill rates over time. We only had a snapshot of MCP-OD fill rates, and only for three MCP-ODs, because these units were stood up during our study. However, the snapshot of MCP-OD fill rates during September 2016 showed which MCP-OD positions, MOSs/AOCs, and grades had low fill rates at that point and which ones were filled by personnel with MOSs/AOCs and/or grades that did not match what was authorized for the MCP-OD positions.[116]

As of September 2016, some MCP-ODs had been stood up, while others had not. We focused on a few that had most or all of the total authorized number of authorized personnel and looked at how those personnel were distributed among MOSs/AOCs. Tables 4.2, 4.3, and

[115] One indicator of personnel readiness is an available strength or fill rate. The highest personnel readiness level is P-1. A unit MOS/AOC needs to have a personnel fill rate (available strength) of 90–100 percent to be at level P-1 (Headquarters, Department of the Army, Army Regulation 220-1, Army Unit Status Reporting and Force Registration—Consolidated Policies, April 15, 2010, p. 44). To identify MOSs/AOCs with readiness issues—that is, those at level P-2, P-3, or P-4—we looked at those with less than 90-percent fill.

[116] Appendix B contains tables summarizing fill rate patterns in division HQ over time. We identified MOSs/AOCs with personnel fill challenges that include average fill rates below 90 percent over time and fluctuating fill rates during recent fiscal years. (To identify fluctuating fill rates during fiscal years, we looked at the coefficient of variation—standard deviation of monthly fill rates for an MOS/AOC divided by the average fill rate for the MOS/AOC.)

4.4 show the fill rates, by MOS/AOC, for the MCP-OD #1, MCP-OD #2, and MCP-OD #3, respectively. In MCP-OD #1, about 68 out of 91 personnel were assigned. Shortfalls were largely in MI specialties, specifically, intelligence analysts (MOS 35F), imagery analysts (35G), human intelligence collectors (35M), and signal intelligence analysts (35N). Public affairs (46Q, 46Z) was also an area with shortfalls. Also, about 28 out of 68 personnel had an MOS/AOC that did not match the position, and about 17 of those had MOS 09B, indicating a trainee not yet assigned an MOS.

MCP-OD #2 had more than its authorized number of personnel (95 assigned, 91 authorized). This MCP-OD had more than enough personnel in some specialties but had shortfalls in many of the same areas as the MCP-OD #1 (e.g., military intelligence, public affairs, budget). Also, about 39 out of 95 personnel had an MOS/AOC that did not match the position, and about 29 of those had MOS 09B.

Unlike the other two MCP-ODs, MCP-OD #3 had at least 100 percent fill in all its positions except the budget officer (36A) position. But of its 92 personnel assigned, 15 had an MOS that did not match the authorized MOS for their position, and 8 of those 15 were MI positions filled by personnel with MOSs such as 11B (infantryman), 92A (automated logistics specialist), and 92Y (unit supply specialist).

Thus, some MCP-ODs have more available strength than others, and, so far, manning shortfalls tend to be largely in MI specialties. In a few cases, obtaining the required clearances were mentioned as a challenge. Because these are new units, the fact of shortfalls is not surprising. What we find important is the pattern in the shortages. We hypothesize that whatever National Guard recruiting or training challenges make it hard for units to initially fill certain positions will arguably make it hard to replace the initial incumbents in the future.

While MOS shortfalls are typically considered a limitation, some interview subjects offered an alternative perspective, highlighting the skills outside of MOS training that can help one succeed as a staff officer. Many AC interviewees had positive comments about the skills and other characteristics that MCP-OD personnel brought to the units. MCP-OD personnel contributed transferable and "real-life" skills from their civilian jobs, such as vehicle maintenance, computer skills, and operations-center management, that enhance the mission.[117] An interview participant commented that ARNG soldiers had multiple MOSs, so they could be cross-trained and move around (on the staff). MCP-OD personnel were also viewed as bringing other attributes to the job, including being more motivated, proactive, and mature compared to AC soldiers, in part, because of their life experiences outside of the Army.[118] One interviewee reported that they could leverage highly motivated and professional ARNG soldiers even if they did not match specified MOS requirements, and another reported after deployment that they left some AC soldiers at home station in favor of RC soldiers who were really good.[119] One senior AC leader referred to the "intellectual diversity" brought by the RC soldiers.[120]

Tables 4.2 through 4.4 depict the fill rates for a sample of three MCP-ODs.

---

[117] AC field-grade officer, Ft. Campbell, Ky., January 10, 2017; AC company-grade officer, Ft. Bliss, Tex., February 14, 2017.

[118] AC field-grade officer, Ft. Campbell, Ky., January 10, 2017; Ft. Campbell, January 9, 2017; AC G2 officer, Ft. Campbell, Ky., January 9, 2017; ARNG field-grade officer, Ft. Bliss, Tex., February 12, 2017; AC NCO, Ft. Bliss, Tex., February 13, 2017; AC field-grade officer, Ft. Bragg, N.C., April 10, 2017.

[119] AC officer, Ft. Campbell, Ky., January 9, 2017.

[120] ARNG field-grade officer, Ft. Bragg, N.C., June 13, 2017.

Table 4.2
Fill Rates by MOS/AOC for Sample MCP-OD #1

MOS/ AOC	Position Title	Authorized Quantity	On-hand Quantity as of 9/16	Percent Fill	Number Filled with Lower Grade	Number Filled with Different MOS	Number Filled with MOS/ AOC 09B (Trainee)
01A00	Operations officer	2	2	100	1		
02A00	Liaison officer	7	6	86			
11A00	Plans officer	1		0			
11B2P	Assistant operations sergeant	1	1	100			
11B3P	Operations sergeant	1	1	100			
11Z5P	First sergeant	1	1	100			
12B3P	Operations sergeant	1	1	100			
12B4P	Operations sergeant	1	1	100			
12Y1P	Geospatial engineer	1		0			
12Z5P	Senior engineer NCO	1	1	100			
13A00	Assistant fire support officer	1	1	100			
13F4P	Fire support NCO	1	1	100			
13J2P	Fire control sergeant	1	1	100		1	1
14G1P	Operations assistant	1	1	100	1	1	1
14G2P	Assistant operations sergeant	1	1	100		1	
14G3P	Operations sergeant	1	1	100			
153AP	TAC operations officer	1	1	100			
15P3P	Assistant aviation OPS sergeant	1		0			
19Z6P	Operations sergeant	1	1	100		1	
25B2P	Senior information tech specialist	1		0			
25L1P	Cable SYS INSTL-MNT	3	3	100	2	1	1
25L3P	Cable SYS team chief	1	1	100	1	1	
25U1P	Signal support SYS SP	1	1	100	1	1	1
27A00	Team chief	1		0			
27D3P	OPS law NCO	1	1	100	1		
30A00	IO assessment	1	1	100		1	
31B5P	Operations NCO	1		0			
35D00	Operations officer	2	2	100	1		
35F1P	Intel analyst	8	4	50	3	3	3
35F2P	Intel analyst	5	4	80	1	1	1

Table 4.2—Continued

MOS/ AOC	Position Title	Authorized Quantity	On-hand Quantity as of 9/16	Percent Fill	Number Filled with Lower Grade	Number Filled with Different MOS	Number Filled with MOS/ AOC 09B (Trainee)
35F3P	ASAS master analyst	1	1	100		1	
35F4P	Senior intel sergeant	1	1	100	1		
35G1P	Imagery analyst	4	1	25			
35G2P	Imagery analyst	1	1	100	1		
35G3P	Imagery sergeant	1	1	100		1	
35M1P	HUMINT collector	3	2	67		2	
35N1P	SIGINT analyst	4	3	75	1	3	2
35N2P	SIGINT analyst	1	1	100			
35T1P	MI SYS MNTR/INTGR	1	1	100	1	1	1
36A00	Budget officer	1		0			
36B3P	SR financial analyst	1	1	100	1		
37A00	PSYOP officer	0		N/A			
37F4P	Senior PSYOP sergeant	0		N/A			
38A00	Civil affairs officer	0		N/A			
38B4P	Civil affairs NCO	0		N/A			
38B5P	CA operations sergeant	0		N/A			
42A1P	Human resources SPC	5	6	120	3	5	3
42A2P	Human resources sergeant	1	1	100			
46Q2P	Public affairs sergeant	1		0			
46Z4P	Public affairs OPS NCO	1		0			
57A00	Mission command integrator	1		0			
74D1P	CBRN specialist	1	1	100	1		
74D2P	CBRN NCO	2	2	100	2	1	
90A88	Trans OPS officer	1	1	100			
90A92	Fuel operations officer	1		0			
91B1P	Wheeled vehicle mechanic	1	1	100			
91B2P	Wheeled vehicle mechanic	1	1	100			
91C1P	Utilities equipment rep	1	1	100			
91D1P	TAC power generator SPC	1	1	100	1	1	1
91X4P	Army maintenance sergeant	1		0			
92Y1P	Supply specialist	1	1	100	1	1	1
92Y3P	Supply sergeant	1	1	100		1	

SOURCE: RAND Arroyo Center analysis of FMSWeb data.

KEY: ASAS = All Source Analysis System ; CBRN = chemical, biological, radiological, nuclear; HUMINT = human intelligence; INSTL-MNT = installer-maintainer; IO = information operations; MI SYS MNTR/ INTGR = Military Intelligence Systems Maintainer/Integrator; PSYOP = psychological operations; SP/SPC = specialist; SYS = system; Trans OPS = transportation operations.

Table 4.3
Fill Rates by MOS/AOC for MCP-OD #2

MOS/ AOC	Position Title	Authorized Quantity	On-hand Quantity as of 9/16	Percent Fill	Number Filled with Lower Grade	Number Filled with Different MOS	Number Filled with MOS/ AOC 09B (Trainee)
01A00	Operations officer	2	1	50			
02A00	Liaison officer	7	7	100	2	5	
11A00	Plans officer	1	1	100			
11B2O	Assistant operations sergeant	1	3	100			
11B3O	Operations sergeant	1	1	100			
11Z5M	First sergeant	1	1	100			
12B3O	Operations sergeant	1	1	100	1	1	
12B4O	Operations sergeant	1		0			
12Y1O	Geospatial engineer	1	1	100			
12Z5O	Senior engineer NCO	1	1	100			
13A00	Assistant fire support officer	1		0			
13F4O	Fire support NCO	1	1	100			
13J2O	Fire control sergeant	1	1	100			
14G1O	Operations assistant	1	2	200	2	1	1
14G2O	Assistant operations sergeant	1	1	100		1	1
14G3O	Operations sergeant	1		0			
153AI	TAC operations officer	1	1	100			
15P3O	Assistant aviation OPS sergeant	1		0			
19Z6O	Operations sergeant	1	1	100		1	
25B2O	Senior information tech specialist	1	2	200			
25L1O	Cable SYS INSTL-MNT	3	5	167	4	4	4
25L3O	Cable SYS team chief	1	1	100			
25U1O	Signal support SYS SP	1	2	200	2	2	2
27A00	Team chief	1	1	100			
27D3O	Operations law NCO	1		0			
30A00	IO assessment	1		0			
31B5O	Operations NCO	1	1	100			
35D00	Operations officer	2	1	50			
35F1O	Intel analyst	8	8	100	2	5	4
35F2O	Intel analyst	5	2	40			

Table 4.3—Continued

MOS/ AOC	Position Title	Authorized Quantity	On-hand Quantity as of 9/16	Percent Fill	Number Filled with Lower Grade	Number Filled with Different MOS	Number Filled with MOS/ AOC 09B (Trainee)
35F3O	ASAS master analyst	1	3	300		2	
35F4O	Senior intel sergeant	1		0			
35G1O	Imagery analyst	4	3	75	3	2	2
35G2O	Imagery analyst	1		0			
35G3O	Imagery sergeant	1	2	200		1	
35M1O	HUMINT collector	3	2	67	1	1	1
35N1O	SIGINT analyst	4	3	75	1	2	2
35N2O	SIGINT analyst	1	1	100	1		
35T1O	MI SYS MNTR/INTGR	1	1	100		1	1
36A00	Budget officer	1		0			
36B3O	SR FIN analyst	1	1	100	1		
37A00	PSYOP officer	0					
37F4O	Senior PSYOP sergeant	0					
38A00	Civil affairs officer	0					
38B4O	Civil affairs NCO	0					
38B5O	CA operations sergeant	0					
42A1O	Human resources SPC	5	11	220	2	7	5
42A2O	Human resources sergeant	1	2	200			
46Q2O	Public affairs sergeant	1	1	100	1	1	1
46Z4O	Public affairs OPS NCO	1		0			
57A00	Mission command INTEGR	1		0			
74D1O	CBRN specialist	1	2	200	2	2	2
74D2O	CBRN NCO	2	3	150	1	1	
90A88	Trans OPS officer	1	1	100	1	1	
90A92	Fuel operations officer	1	2	200	1	1	
91B1O	Wheeled veh mech	1	2	200			
91B2O	Wheeled veh mech	1	1	100			
91C1O	Utilities equipment rep	1	1	100	1		
91D1O	TAC PWR gen spec	1	2	200	1	2	2
91X4O	Army maintenance sergeant	1		0			
92Y1O	Supply specialist	1	2	200	1	1	1
92Y3O	Supply sergeant	1	2	200		1	

SOURCE: RAND Arroyo Center analysis of FMSWeb data.

Table 4.4
Fill Rates by MOS/AOC for MCP-OD #3

MOS/ AOC	Position Title	Authorized Quantity	On-hand Quantity as of 9/16	Percent Fill	Number Filled with Lower Grade	Number Filled with Different MOS	Number Filled with MOS/ AOC 09B (Trainee)
01A00	Operations officer	2	3	150			
02A00	Liaison officer	7	7	100	2		
11A00	Plans officer	1	1	100	1		
11B2O	Assistant operations sergeant	1	1	100			
11B3O	Operations sergeant	1	1	100	1		
11Z5M	First sergeant	1	1	100			
12B3O	Operations sergeant	1	1	100			
12B4O	Operations sergeant	1	1	100		1	
12Y1O	Geospatial engineer	1	1	100			
12Z5O	Senior engineer NCO	1	1	100	1		
13A00	Assistant fire support officer	1	1	100	1		
13F4O	Fire support NCO	1	1	100			
13J2O	Fire control sergeant	1	1	100			
14G1O	Operations assistant	1	1	100			
14G2O	Assistant operations sergeant	1	1	100		1	
14G3O	Operations sergeant	1	1	100			
153AI	TAC operations officer	1	1	100			
15P3O	Assistant aviation OPS sergeant	1	1	100			
19Z6O	Operations sergeant	1	1	100		1	
25B2O	Senior information tech SP	1	1	100			
25L1O	Cable SYS INSTL-MNT	3	3	100			
25L3O	Cable SYS team chief	1	1	100			
25U1O	Signal support SYS SP	1	1	100			
27A00	Team chief	1	1	100			
27D3O	Operations law NCO	1	1	100			
30A00	IO assessment	1	1	100		1	
31B5O	Operations NCO	1	1	100	1		
35D00	Operations officer	2	2	100	1		
35F1O	Intel analyst	8	8	100			
35F2O	Intel analyst	5	5	100			
35F3O	ASAS master analyst	1	1	100			
35F4O	Senior intel sergeant	1	2	200	1		

Table 4.4—Continued

MOS/ AOC	Position Title	Authorized Quantity	On-hand Quantity as of 9/16	Percent Fill	Number Filled with Lower Grade	Number Filled with Different MOS	Number Filled with MOS/ AOC 09B (Trainee)
35G1O	Imagery analyst	4	4	100		4	
35G2O	Imagery analyst	1	1	100			
35G3O	Imagery sergeant	1	1	100			
35M1O	HUMINT collector	3	3	100		1	
35N1O	SIGINT analyst	4	4	100		2	
35N2O	SIGINT analyst	1	1	100			
35T1O	MI SYS MNTR/INTGR	1	1	100		1	
36A00	Budget officer	1		0			
36B3O	SR FIN analyst	1	1	100			
37A00	PSYOP officer	0					
37F4O	Senior PSYOP sergeant	0					
38A00	Civil affairs officer	0					
38B4O	Civil affairs NCO	0					
38B5O	CA operations sergeant	0					
42A1O	Human resources SPC	5	5	100			
42A2O	Human resources sergeant	1	1	100			
46Q2O	Public affairs sergeant	1	1	100	1		
46Z4O	Public affairs OPS NCO	1	1	100	1		
57A00	Mission command INTEGR	1	1	100	1	1	
74D1O	CBRN specialist	1	1	100			
74D2O	CBRN NCO	2	2	100			
90A88	Trans OPS officer	1	1	100	1	1	
90A92	Fuel operations officer	1	1	100	1	1	
91B1O	Wheeled veh mech	1	1	100			
91B2O	Wheeled veh mech	1	1	100			
91C1O	Utilities equipment rep	1	1	100			
91D1O	TAC PWR gen spec	1	1	100			
91X4O	Army maintenance sergeant	1	1	100			
92Y1O	Supply specialist	1	1	100			
92Y3O	Supply sergeant	1	2	100	1		

SOURCE: RAND Arroyo Center analysis of FMSWeb data.

Supporting Personnel in a Consolidated Unit

While many issues with the FARG II design are naturally resolved in the course of a deployment (soldiers get familiar with their assigned systems: teams "form, norm, and storm"), one aspect may get harder with time. Until mobilized and transferred to AC control, ARNG soldiers are assigned to ARNG units and have their state HR infrastructure to manage their personnel actions. When a deployed ARNG soldier needs a personnel action completed, an RA soldier in the HHBn or other office may not have access to the appropriate programs to help. An RC HR person would be unable to support an RA soldier's request until given access to and training in the AC HR systems.[121] Interview subjects noted that this is not only a matter of objective effectiveness in completing assigned tasks; the AC-RC nature of the problem can generate dissatisfaction and tension within the HQ if the ARNG soldiers feel their support needs are not being met.[122]

Facilities

"Facilities analysis examines military property, installations and industrial facilities (e.g. government owned ammunition production facilities) that support our forces to see if they can be used to fill in a capability gap. Is there a lack of operations and maintenance? Is the problem caused, at least in part, by inadequate infrastructure?"[123]

As in the discussion of materiel, this section asks whether the elements of the division HQ have the broader infrastructure needed to attain and sustain readiness. Because we are discussing units comprising multiple components, the answer must include distance, as well as the facilities themselves. Because facilities for the AC portion of the structure were rarely mentioned as an issue, we will focus on where the states have stationed their part of the structure.

We also try to make explicit the connection between this and other functions. For example, an excellent ARNG center[124] may be located too far from the ideal population for effectively recruiting MCP-OD personnel, or it may be located close to the recruiting base but so far from the division's installation that training opportunities with the AC become rarer or more expensive.

Main Command Post–Operational Detachments Stationing

One factor that makes it difficult to compare MCP-ODs and develop mitigating strategies for their challenges is the vast differences in their geographic relationship with their supported division HQ. Table 4.5 lists the MCP-ODs in ascending order of their distance from the AC installation.

---

[121] This is not limited to MCP-ODs, of course. Any time individuals of one component augment a unit from another, this problem is likely to occur to some degree.

[122] AC NCO, Ft. Campbell, Ky., January 10, 2017.

[123] AcqNotes, "JCIDS Process: DOTMLPF-P Analysis."

[124] Or USAR center, in the case of the MCP-OD for the 25th ID.

Table 4.5
Distances Between MCP-OD Home Station and Partner Division HQ

Division	Division Home Station	MCP-OD Source	MCP-OD City	Distance to MCP-OD
3rd ID	Ft. Stewart, Ga.	Georgia Army National Guard	Ft. Stewart, Ga.	0
25th ID	Schofield Barracks, Hawaii	US Army Reserve	Schofield Barracks, Hawaii	0
1st Armor Division	Ft. Bliss, Tex.	Texas Army National Guard	El Paso, Tex.	7
1st Cavalry Division	Ft. Hood, Tex.	Texas Army National Guard	Round Rock, Tex.	50
10th Mountain	Ft. Drum, N.Y.	New York Army National Guard	Auburn, N.Y.	112
82nd Airborne	Ft. Bragg, N.C.	North Carolina Army National Guard	Charlotte, N.C.	124
1st ID	Ft. Riley, Kan.	Nebraska Army National Guard	Lincoln, Neb.	152
101st Airborne	Ft. Campbell, Ky.	Kentucky Army National Guard	Louisville, Ky.	190
4th ID	Ft. Carson, Colo.	Utah Army National Guard	Camp Williams, Utah	577

NOTE: Distances using Google Maps from city center to installation center, September 15, 2017.

Interviewees from both the AC and RC evinced a strong consensus that colocating the MCP-OD with its partner division HQ was the optimal stationing solution.[125] While the best location might vary according to the geographic dispersion of MCP-OD personnel within a state, there are distinct advantages of colocation. In case the MCP-OD is mobilized for a mission employing a home station command post, colocation would present fewer transition challenges compared to being located elsewhere. The integration of training with division HQ personnel would also be better enabled during IDT periods that align with AC soldier availability, as well as providing access to division facilities and computer networks.[126]

The effect of such stationing on recruiting, however, might be a two-edged sword. On one hand, locating the MCP-OD on or near the division installation might facilitate the recruiting of individuals who either separate from active duty after serving at the unit or those with civilian jobs that bring them to the area. In either case, they might have experience working with the division that could make their integration into the MCP-OD and the division HQ easier than for most ARNG members. On the other hand, many AC divisions are located away from major population centers in their states. Thus, it might be harder to recruit existing ARNG soldiers to serve in the MCP-OD, and MCP-OD members might find it hard to move to ARNG positions in other parts of the state.

[125] ARNG field-grade officer, Ft. Bliss, Tex., February 12, 2017; AC NCO, Ft. Bliss, Tex., February 13, 2017. AC NCO, Ft. Bliss, Tex., February 13, 2017.

[126] Two division staff officers, Schofield Barracks, Hawaii, April 10, 2017.

This situation is further complicated by the fact that a MCP-OD must be supported by a single state, at least prior to activation. Note that two of the eight ARNG-provided MCP-ODs come from a state other than the one hosting the AC installation, and a third base, Fort Campbell, sits astride the state boundary.

In the case of the MCP-OD for the 82nd Airborne, interviewees uniformly stated that the opportunity to be in an airborne unit, and particularly to be part of the 82nd Airborne, generally overcame the disadvantages of Fort Bragg being far removed from the state's largest cities, and recruiting for an ARNG unit stationed there was not difficult. However, this might not hold true over time or for other divisions. Additionally, the value of colocation is related to the value of premobilization collective training. If the vast majority of collective training occurs after mobilization, colocation may be less important.

Because of these conflicting potential effects and the unique features of each MCP-OD's location, this report does not offer any recommendations for mitigation strategies. Over time, the state ARNG leadership and the AC divisions will learn what effects predominate in their case and whether a change in the MCP-OD's location is likely to provide net positive effects on manning and training of the unit.

Summary of DOTMLPF-P Issues

In this chapter, we set out to use the DOTMLPF-P framework to organize our observations from both interviews and analysis of Army databases. We have highlighted issues relating to each function, and for each we discuss its source, how we attempted to measure or assess it, and the possible ways to mitigate any negative effects it may have on the consolidated division HQ's readiness to deploy.

None of these DOTMLPF-P issues are "critical," in the sense of rendering the division HQ unable to accomplish its mission. However, each of them presents some hurdle that the HQ must overcome if it is to maximize its readiness for operations across the range of contingencies. The most critical of these relate to how RC soldiers are brought into the MCP-OD and then given the individual and collective training to perform their duties. Most of the other functional issues in some way exacerbate these two fundamental challenges. The following chapter builds on this analysis and attempts to show how these incremental effects combine to affect the net readiness of the HQ in various situations.

CHAPTER FIVE

Impact of Focus Area Review Group II Changes on Short-Notice Readiness

While the above analysis identifies several specific challenges to the readiness of the new, consolidated division HQ, it does not predict the net effectiveness of the FARG II HQ when it deploys. As a preliminary assessment of this aspect, we created a model, shown in Figure 5.1, that uses various characteristics about contingency readiness at the moment of alert and likely scenarios requiring HQ deployment. For each scenario, we estimated the HQ's capacity to execute mission-essential tasks broken down by warfighter function.[1] Note that all these assume none of the above mitigation policies have been systematically implemented. While available data did not support the calibration of the model for quantitative predictive purposes, the model in its current form can give a sense of which areas are likely to be problematic under the various scenarios. Further development of model mechanisms and calibration against data would be required to transform the model into a useful planning tool.

Model Description

The model uses a system dynamics approach to represent the staffing dynamics of the MCP-OD and its integration with the AC elements of the HQ. It can be divided into four sectors: MCP-OD Staffing Dynamics, AC Fill Dynamics, Mission Staffing Needs, and HQ Warfighter Function Effectiveness. The layout of each of these sectors and the interlinkages between them are shown in Figure 5.1. The model handles staffing for each warfighter function separately through the use of vectored variables. In the model diagram, these variables are represented by stacked symbols. Each of the sectors will be described in turn.

The MCP-OD Staffing Dynamics sector traces the staffing of the MCP-OD at a high level of abstraction. Positions transition from being authorized on the MCP-OD MTOE, through the process of manning, to become MCP-OD manned positions. Different scenarios assume different levels of initial manning at the time of notice. Once they fill the positions, the soldiers in them may require additional individual training. Once they have completed the individual training process, they transition to the MCP-OD trained state. MCP-OD trained soldiers become integrated into the HQ through the process of collective training, and with

[1] As discussed earlier, necessary capability is having at least one AC soldier for each skill set required in the various sections, cells, boards, and working groups within each command post. Capacity is measured by how many shifts contained the required skill sets.

Figure 5.1
Headquarters Staffing Model Diagram

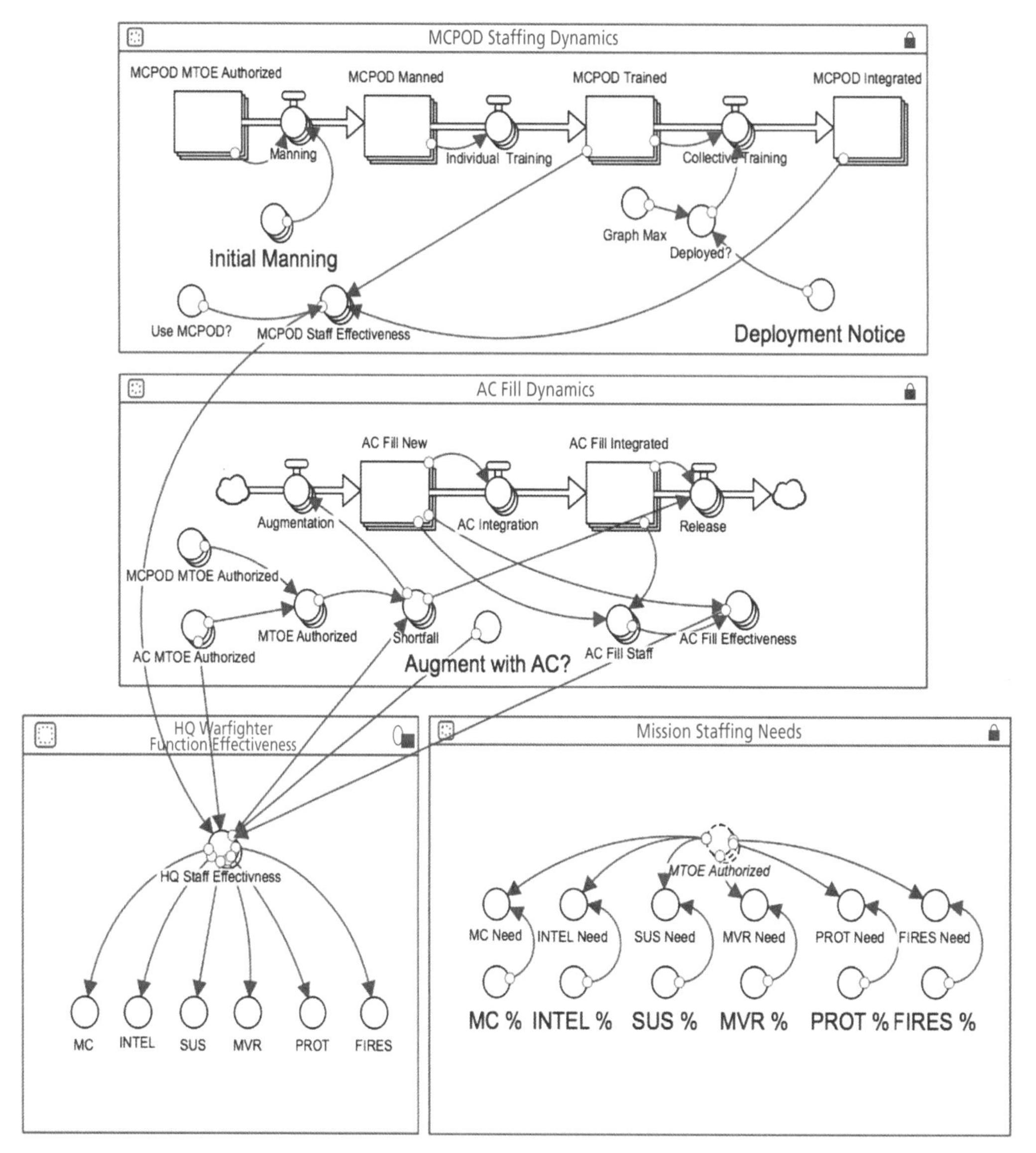

SOURCE: RAND Arroyo Center analysis.
KEY: INTEL = intelligence; SUS = sustainment; MVR = maneuver; PROT = protection.

that complete, they transition into the state of being MCP-OD integrated. The MCP-OD can deploy with soldiers who have completed their individual but not their collective training but is less effective under these conditions. Deployment, however, serves as a kind of collective training, producing fully integrated MCP-OD soldiers after a few weeks in the field. The MCP-OD Staffing Dynamics sector includes several parameters for reflecting different scenarios, including the level of initial manning (expressed as a percent of MTOE authorized positions filled), the number of days of notice before deployment, and whether the MCP-OD is to be deployed. MCP-OD MTOE authorizations for positions associated with each warfighter function are based on data from the 101st Airborne.

The AC Fill Dynamics sector presents a notional representation of how soldiers can be pulled from other duties to augment HQ functions when RC component soldiers cannot be made available quickly enough. This sector is driven by staffing shortfalls. If the decision is made to augment with AC soldiers, empty spots are filled through the process of augmentation, which takes about a week on average. Once added through augmentation, soldiers are in the AC fill new state, where they can be deployed but are not fully effective. They are integrated in the HQ through the process of AC integration, which also takes about a week on average, at which point they are fully effective. As spots are filled from the MCP-OD, soldiers are released back to their other duties, with some overlap to bring the new MCP-OD soldier up to speed.

The HQ Warfighter Function Effectiveness sector sums the effectiveness of the MCP-OD and AC fill parts and breaks them out by warfighter function for graphical output. It also provides feedback on shortfall to the AC Fill Dynamics sector.

Finally, the Mission Staffing Needs sector allows scenarios to be set up by specifying the percent strength required in the HQ unit for each of the six warfighter functions. These are compared to the MTOE authorization to find the number of soldiers needed in each function and the result is plotted on the graphical output as a dotted line.

Parameters

The model, as it currently stands, presents several parameters that can be adjusted to reflect different assumptions and scenarios. These include the following:

- mission staffing requirements for MCP-OD (by warfighter function)
- initial manning levels of MCP-OD (by warfighter function, as percent of MTOE authorization)
- MTOE authorization levels (by warfighter function, both RC and AC)
- notice of deployment, mobilization period, and deployment date
- rates of manning, individual training, and collective training
- AC augmentation (yes or no)
- rate of AC augmentation
- rate of AC integration.

Modeling Assumptions

In constructing these models, we made the following assumptions:

1. Only the soldiers assigned to the main command post are modeled. Few MCP-OD soldiers are assigned to the HHBn, DTAC, or the SACP of a division HQ, and our observations indicated these numbers vary between divisions. MTOE for MCP-OD MCP staff is 88, while MTOE for AC MCP staff is 245, for a total of 333 personnel authorized.
2. ARNG soldiers who are already assigned to the MCP-OD can be notified of an upcoming deployment within a few days.

3. To be fully ready for deployment, the average ARNG soldier requires some additional individual and collective training after notification. MCP-OD positions, even when already filled at the time of notice, are not assumed to be fully DMOSQ without some additional training and preparation.
4. Most (but not all) ARNG soldiers assigned to the MCP-OD can complete their individual training and readiness requirement within 30 days of notification.
5. Most ARNG soldiers can complete collective training and integration into the MCP within 90 days in the predeployment environment. If time does not permit this integration before deployment, however, most individually qualified ARNG soldiers will be integrated through on-the-job training within 15 days after deployment.
6. Replacements for non-DMOSQ soldiers are sourced directly from Regular Army soldiers on the division's home installation. (The ARNG is supposed to source DMOSQ replacements first and only rely on the AC when it cannot meet manpower needs. However, the ARNG replacement process takes time, and anecdotal evidence from interviews showed that, for quick mobilization requirements, the AC will often short-circuit the ARNG process and replace non-DMOSQ ARNG soldiers with AC soldiers.)
7. AC replacements take an average of one week to be found, receive orders, and join the HQ.
8. Collective training for AC replacements is completed one week *after* they report to the MCP through on-the-job training. Therefore, full readiness, inclusive of collective training, does not happen until an average of one week after reporting (two weeks after notification).
9. The full MTOE for the MCP-OD MCP includes five positions to be filled by the USAR. For the sake of simplicity, these positions are not treated in this model. It is assumed that these positions will either be easily filled from USAR forces or will be quickly filled with AC replacements. A full treatment of MCP-OD staffing would need to include USAR dynamics and staffing requirements.

To make the model output comprehensible, we grouped MCP soldiers by warfighter function. MCP-OD positions were assigned to primary warfighter functions based on their paragraphs in the MTOE. In cases where the warfighter function that an MOS would typically serve was mixed or ambiguous, they were grouped into the most likely warfighter function.[2] A complete list of MCP-OD positions from a sample division HQ grouped by warfighting function is found in Table 5.1 below. The sample is used for illustrative purposes; however, all AC division HQ have identical FARG II MTOEs.

Scenarios for Headquarters Deployment

The following is an unclassified set of scenarios for deployment of a MCP-OD under various mission requirements and various initial DMOSQ fill levels. While not a comprehensive list of all possible missions, it provides a plausible range of contingencies for which a division HQ

---

[2] We accept that this is open to interpretation. For example, the Engagement and Information Operations cells may be seen as Mission Command elements rather than part of the Maneuver staff. We count them in the latter based on their place in the MTOE structure.

Table 5.1
MCP-OD Grouped by Warfighting Function

Section	Fires	INTEL	MC	MVR	PROT	SUS	Total
Company HQ			7				7
MCP/COFS/CMD LNO SEC			7				7
MCP/COFS/KM SEC			2				2
MCP/FIRES/AMD SEC	3						3
MCP/FIRES/CUOPS	1						1
MCP/FIRES/FS SEC	2						2
MCP/INTEL		3					3
MCP/INTEL/ACE		18					18
MCP/INTEL/CUOPS		4					4
MCP/INTEL/G2X		5					5
MCP/INTEL/OPS SEC		2					2
MCP/MAINT SEC						4	4
MCP/MVR/CUOPS				6			6
MCP/MVR/ENGMTS				0			0
MCP/MVR/IO SEC				1			1
MCP/MVR/PLANS CELL				2			2
MCP/PROT/CUOPS					1		1
MCP/PROT/PR SECT					1		1
MCP/SIG/CABLE ELM			4				4
MCP/SS/PAO SEC			2				2
MCP/SS/SJA			2				2
MCP/SUS/CUOPS						1	1
MCP/SUS/HR						6	6
MCP/SUS/LOG						2	2
MCP/SUS/RM						2	2
Grand total	6	32	24	9	2	15	88

SOURCE: RAND Arroyo Center analysis of division MTOE data.

KEY: CELL = team; CMD = command; COFS = chief of staff; FS = fire support; KM = knowledge management; LNO = liaison officer; LOG = logistics; MVR = maneuver; OPS SEC = operations section ; PR SECT = personnel recovery section; PROT = protection; RM = resource management; SECT = section; SIG = signals; SS = special staff; SUS = sustainment.

may have to provide mission command. These MCP-OD scenarios include a planned rotational deployment that sets a baseline, a GRF mission, a major contingency (with and without augmentation from the AC), and a short-term humanitarian assistance and disaster relief (HADR) mission.

The use of scenarios is important because a division HQ often responds to a requirement with less than its full authorized strength. This shortfall can be due to a force cap imposed by the combatant commander or higher, a desire to retain the capability for planning and other functions at home station, or simply the expectation that certain skills will not be required to accomplish the mission. For each scenario, the research team estimated the percentage of personnel involved in each warfighting function who would be required for the initial HQ deployment. Because force requirements for recent deployments are generally still classified, as are military plans for future contingencies, the following estimated requirements should be interpreted as illustrative levels derived from the team's experience and interpretation of history and doctrine, not predictions of actual requirements for any specific scenarios. The model itself is as accurate as the parameters set by the operator.

Using the above assumptions and MTOE data, the model was calibrated to show several representative scenarios and their readiness results. In each model, the graph is the central feature that shows MCP-OD readiness by warfighter function. The solid line depicts build-up of actual readiness, while the dotted line is the target readiness. The vertical gray line represents the deployment date (this appears off the graph, to the right, in the first scenario). When the readiness line crosses above its target readiness line before the deployment date, that warfighter function in the MCP-OD is ready as needed. If the readiness line crosses the target line after the deployment date, the gap between target and actual readiness shows how great the readiness gap is and how long it will take for that function to become ready after the MCP deploys. Special attention was paid to instances where threshold readiness is achieved, or where all three lines cross at the same time (as in INTEL in the GRF scenario, Figure 5.2), as that identifies one set of minimum standards to achieve this readiness objective. Furthermore, staff officers with such information could look at current metrics and, if, for example, they found their unit was below the threshold requirement for assigned personnel, identify variables where they could "overachieve" to ensure the unit was still able to meet deployment targets.

Figure 5.2
Rotational Deployment

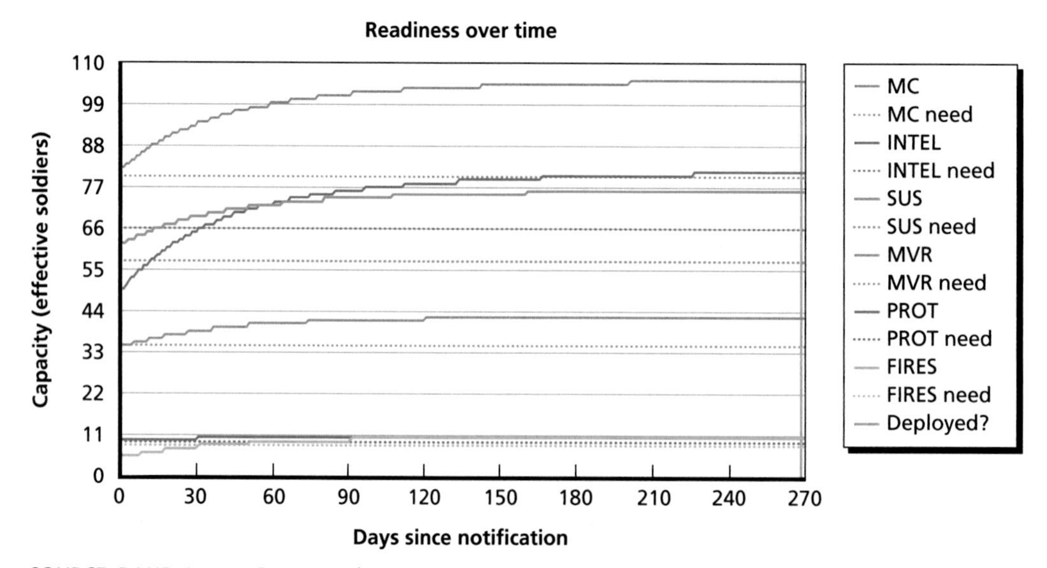

SOURCE: RAND Arroyo Center analysis.

In what follows, we use the model to explore the implications of the MCP-OD architecture for readiness in a series of hypothetical scenarios. In doing this, we remain cognizant that both the scenario specifications and the model dynamics are at best approximations. More work and data will be required to ground the model sufficiently in Army doctrine and practice before this model would be suitable for use as a planning tool.

Notification periods and staffing requirements for each of these types of missions are highly variable depending on the details of the actual missions, so we have selected notional values for key variables that are designed to reflect plausible mission situations while also illustrating key model dynamics. For each scenario, these values (notification period, initial manning of the MCP-OD, and the overall HQ staffing requirement for each warfighter function) are summarized in Table 5.2.

Divided into warfighting functions over time, predicted changes in capacity over time is depicted in the following charts for each of the scenarios. As discussed previously, capability is attained by maintaining at least one trained soldier for each skill set required in the various sections, cells, boards, and working groups within each command post. Capacity is measured by how many shifts contained the required skill sets. This construct is expressed as a percentage on the left side of each chart.

Scenario: Rotational Deployment

The rotational deployment scenario involves a planned deployment where one HQ unit is scheduled to relieve another in the field. There is a full 270 days of notice, and less than full strength is required in each of the warfighter functions. Expressed as a percentage of the total positions on the combined AC and RC MTOEs, the hypothetical strength targets used here are MC, 75 percent (80/106); INTEL, 80 percent (66/82); SUS, 75 percent (58/77); MVR, 80 percent (35/44); PROT, 80 percent (10/12); FIRES, 75 percent (9/12). Overall, this calls for about 77 percent of the MCP-OD and the MCP deploying 258 of its 333 authorized positions. It is assumed that the MCP-OD is manned at 90 percent at the time of notification, though not all the assigned personnel are fully trained and qualified. The relationship of these variables is depicted in Figure 5.2.

Table 5.2
Notional Deployment Scenario Parameters

Scenario	Notification Period (days)	Initial MCP-OD Manning (percent)	Assumed HQ Staffing Requirement (percent)					
			MC	INTEL	SUS	MVR	PROT	FIRES
Rotational	270	90	75	80	75	80	80	75
GRF	30	90	80	80	80	80	80	80
HADR	4	50	80	50	80	70	80	25
Major contingency (No AC fill)	90	80	100	100	100	100	100	100
Major contingency (AC fill)	90	80	100	100	100	100	100	100

The model suggests that the MCP-OD structure presents little risk in this case. Several functions are staffed in the AC at a level sufficient to meet the need: specifically, MC, SUS, and PROT.[3] Other functions come to strength easily within the 270-day notification period. INTEL takes longest to reach adequate strength at 46 days.

Scenario: Global Response Force

The GRF scenario involves a shorter notice period of 30 days. It is assumed that 80-percent strength will be required in all warfighter functions (266 of 333 MTOE authorized positions). Once again, we assume 90-percent manning of the MCP-OD in all warfighter functions at the time of notification. The relationship of these variables is depicted in Figure 5.3.

In the scenario, the model suggests possible problems in two areas: INTEL and FIRES. INTEL reaches the 80-percent strength threshold exactly on the day of deployment. Given the substantial uncertainty involved with this process, this is a less-than-desirable margin for error. FIRES is expected to be about one soldier short until about 45 days after notice.

In this scenario, augmentation from the AC is likely to be needed to assure adequate strength in all areas.

Figure 5.3
Global Response Force Mission Model

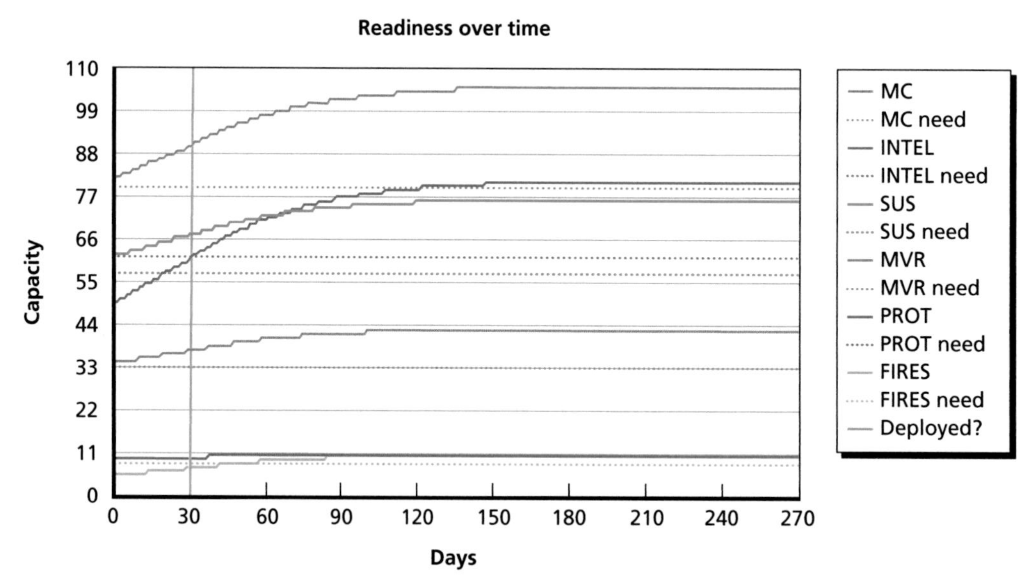

SOURCE: RAND Arroyo Center analysis.

[3] Notification is not the same as mobilization. Thus, notification provides additional time to do the collective training with the division only if the alert order process also provides either additional IDT or AT days, early mobilization, or other means of resourcing additional training days.

Scenario: Short-Term Humanitarian Assistance and Disaster Relief

The short-term HADR scenario involves an extremely short-notice response to a crisis, such as a response to a major earthquake in a partner nation. Because of the lack of anticipation, we assume that the MCP-OD is manned at only 50 percent at the time of notice (44 of 88 positions). Deployment is required four days after notification. Because the mission is quite different from warfighting, different functions of the MCP are required at different levels: MC, 80 percent (85/106); INTEL, 50 percent (40/82); SUS, 80 percent (62/77); MVR, 70 percent (31/44); PROT, 80 percent (10/12); FIRES, 25 percent (3/12). This configuration puts the MCP at an overall 70-percent strength (232/333). The relationship of these variables is depicted in Figure 5.4.

In this scenario, MC has a gap of about one or two soldiers that might be filled by eight days. In this case, limited augmentation from the AC may be required to deploy with adequate capability in all functions.

Scenario: Major Contingency (Without Active Component Augmentation)

In the major contingency scenario, we assume the need for 100-percent strength (333 deployable personnel) in all warfighter functions and 90 days' notice. We further assume that the MCP-OD is manned at only 80 percent at the time of notification. The relationship of these variables is depicted in Figure 5.5.

The MCP-OD structure is not well suited to this contingency. All functions display large gaps at the time of deployment, with MC and INTEL being the shortest staffed. By 140 days, all functions are near adequate strength; however, this is substantially later than the deployment

Figure 5.4
Short-Term Humanitarian Assistance/Disaster Response Mission Model

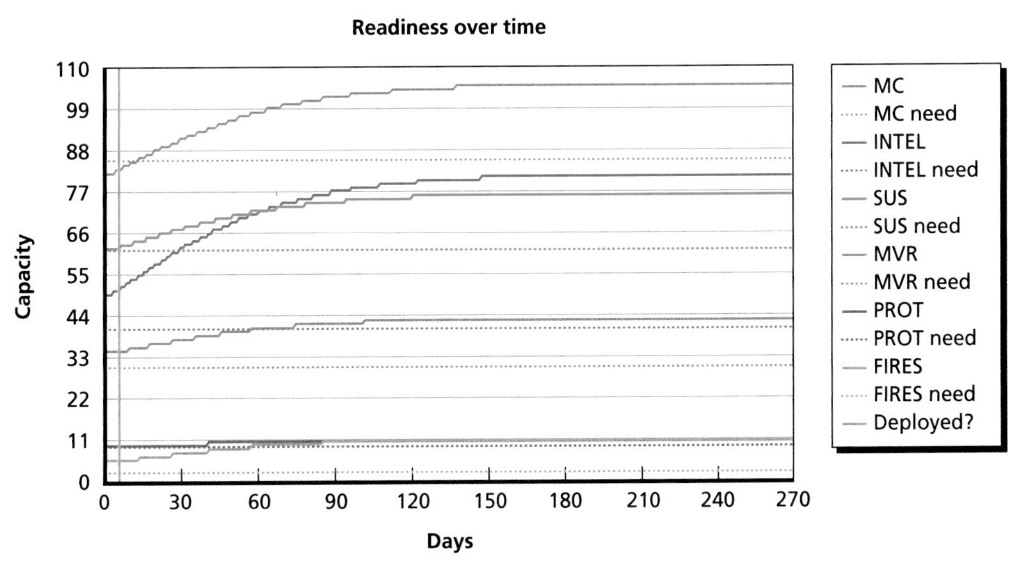

SOURCE: RAND Arroyo Center analysis.

Figure 5.5
Major Contingency Without Active Component Augmentation Model

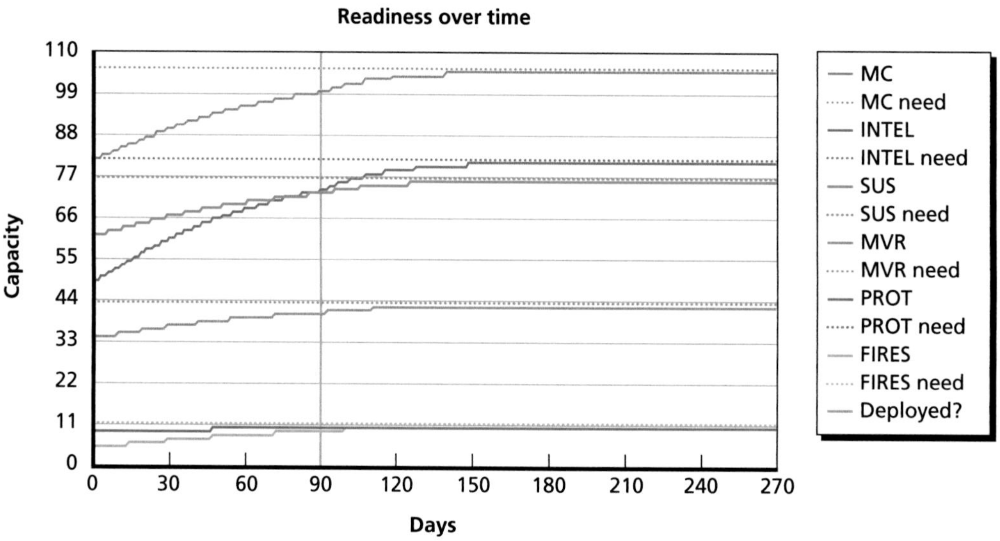

SOURCE: RAND Arroyo Center analysis.

target of 90 days. At 90 days, the MCP is expected to be 24 soldiers short of needed capacity. Specific shortfalls are as follows: MC, 5 short (101/106); INTEL, 6 short (76/82); SUS, 3 short (74/77); MVR, 2 short (42/44); PROT, 1 short (11/12); FIRES, 2 short (10/12).

Scenario: Major Contingency (with Active Component Augmentation)

Anticipating shortfalls, this version of the major contingency scenario involves augmentation from the AC from the time of notification. Again, we assume the need for 100-percent strength in all warfighter functions and 90 days' notice. This time, however we permit augmentation from the active force, a process that can proceed much more quickly than filling and training positions through the RC. As we document in Appendix B, the AC regularly underfills division staff positions in certain low-density MOSs and then fills them as needed to meet deployment requirements. In this case, the model illustrates how this process could be used to staff positions as quickly as possible with AC soldiers and then replace them and return them to their other duties as RC soldiers arrive to take over. The relationship of these variables is depicted in Figure 5.6.

Under the assumptions of the model, AC fill provides a good solution in this scenario. All functions come to full strength within about 15 days and remain at or slightly above full strength until deployment as RC soldiers arrive and overlap with their AC counterparts. While the prior scenario showed a remaining gap of 24 ARNG personnel at the 90-day point, here the model shows considerable flexibility, allowing for as many as 74 AC augmentees at the 20-day point and easily allowing the warfighter functions to reach full capacity well before deployment. Achieving full strength as early as possible (around 20 days after notification) would better enable the staff to conduct the required predeployment planning and preparation and then smoothly transition to a more multicomponent staff, even if some AC personnel

Figure 5.6
Major Contingency with Active Component Augmentation Model

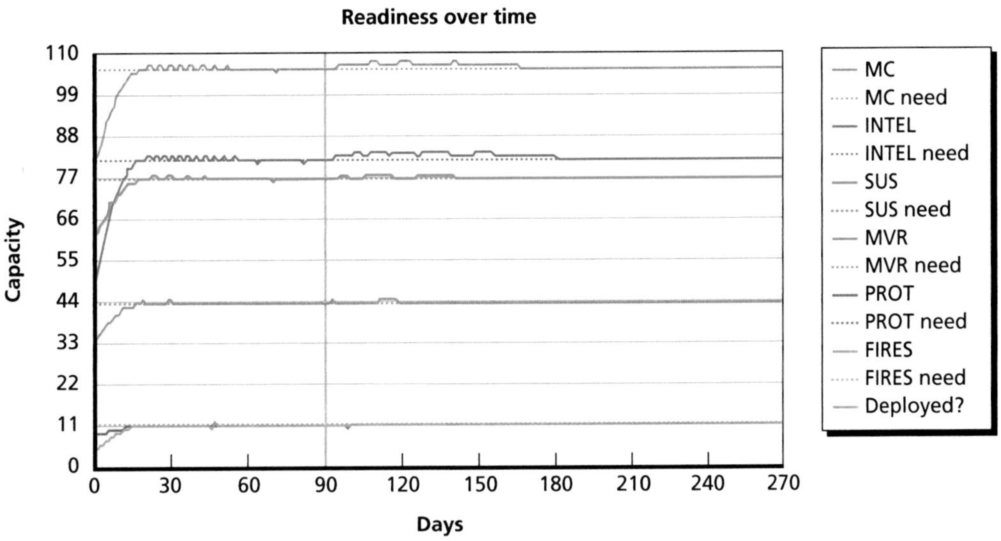

SOURCE: RAND Arroyo Center analysis.

are not required to deploy with the HQ. Under these assumptions, the MCP would actually deploy with 24 AC augmentees filling positions on the ARNG MTOE. The decision to use additional AC augmentees to come to full strength earlier would lie with the appropriate commanders. Of course, this solution is not without costs, as the AC soldiers that are reassigned to the MCP are presumably needed elsewhere, thus leaving gaps in other parts of the division, corps, or installation.

In sum, this model illustrates approximate readiness of the MCP-OD under various conditions and requirements. Under conditions of planned deployments of 270 days' notice, all warfighting groups in a MCP-OD are typically ready. However, under various short-notice conditions, such as GRF, major contingency, or HADR operations, this model reveals portions of the MCP-OD that are not ready by the deployment date. This gap in readiness can be remedied by using AC fill. However, AC fill likely creates stresses elsewhere in the force, as AC personnel become unavailable for other tasks in order to fill seats in the MCP-OD.

Summary

It bears noting that the model is based on each deployment as an independent event. An even more thorough model could factor in the cumulative effects of repeated deployments by the same HQ. Some of these effects would be negative, such as the reduction in deployable personnel immediately following redeployment, until they are either replaced by new soldiers or have had sufficient time to reset their "dwell clock." Other effects could be positive, in that RC personnel who deployed once with the AC HQ and stayed in the MCP-OD would need less pre- and postmobilization training to become fully functioning division-staff members again. The division staff would presumably also get better at anticipating the training and administrative requirements of RC soldiers, making the integration process move more smoothly in each event.

Notwithstanding that caveat, the key finding from use of this model is that the areas in which the FARG II design is most likely to impose risk are those where the Army would be least willing to accept such risks. Positions in the most vulnerable warfighter functions, such as mission command and intelligence, need to be closely monitored as each MCP-OD is standing up. As discussed in the previous chapter, these skills may also be affected by one or more DOTMLPF-P challenges. To the degree manning and training shortfalls persist in these critical skills, the recommended actions in the next chapter will be even more relevant to ensure the full HQ is ready for the most challenging deployment scenarios.

CHAPTER SIX

Conclusion and Recommendations

Conclusion

In the limited number of deployments executed thus far, MCP-ODs have demonstrated that if they are given at least 270 days of advance notice, they can successfully deploy with an AC division HQ and accomplish their missions. However, without substantial advanced notification of sourcing, MCP-OD personnel will not be able to deploy as quickly as the AC soldiers in a division's command posts. This limitation was known when FARG II was designed and explicitly accepted as a risk by the CSA. As noted by two of the people interviewed by the research team, the decision was not the result of "a good idea[;] it was a necessary reaction to resource constraints and the best of a menu of bad choices."[1]

Stakeholders may not like the existence of this risk, but given the imperative to reduce the size of the Army, the cuts to division (and corps) HQ structures were accepted by the Army's leadership as a trade-off to preserve force structure elsewhere. Therefore, the major issues for this study were whether the risk is significantly greater than what was anticipated by FARG II and whether the accepted and the previously unknown risks might be further mitigated.[2]

Although we fleshed out additional detail on the risks, such as the probable time lines for MCP-OD availability, the instances of risks found in our research—to include discussions with numerous stakeholders and multiple echelons—were adequately anticipated by the FARG II designers: the types of mitigation of the FARG I risks that the FARG II design and the MCP-OD design were supposed to address do in fact address them. Nonetheless, there is room for improvement in both design and execution. We found several implications of MCP-OD readiness and availability limitations that might be mitigated. We summarize our key findings and provide recommendations for additional mitigation of risk in the following sections.

---

1 Interview with individuals at the CAC who had been involved with FARG design and implementation, Ft. Leavenworth, Kan., January 4, 2017.

2 As discussed in Chapter Two, the MCP-ODs were themselves an initiative to mitigate the risk of previously directed cuts to division and corps HQ.

Key Findings

Our analysis resulted in the following:

1. For the most part, we validated the risks anticipated by the FARG II designers. However, many of our interlocutors in division HQ were not aware that these risks were known and had been accepted by Army leadership as a trade-off for keeping AC force structure elsewhere. Several interviewees merely restated these risks. However, the risks to division HQ readiness for short-notice deployments may be greater than what was estimated by FARG II because MCP-ODs are partner units rather than MCUs and are not resourced for additional training days as the designers had expected.
2. As the Army had anticipated, MCP-ODs affect the capacity rather than the capabilities of division HQ. Although they have an "AA" UIC, MCP-ODs are more akin to a set of augmentees in reserve rather than a "unit" that operates in a cohesive, collective fashion compared to a maneuver unit or a division HHBn. From this situation, we inferred that other sources of personnel, such as individual augmentation taskings, might also be a means of increasing capacity.
3. For deployments with sufficient advance notice (i.e., rotational or "patch chart" deployments) that allow time to complete training and mobilization requirements, MCP-ODs are likely to be a satisfactory solution to most of the loss of capacity resulting from the reduction in division HQ personnel authorizations. While the model used in Chapter Five is meant to be illustrative, to the extent it accurately captures the basic dynamics and parameters of a HQ deployment, it supports this finding.
4. Under current policy and resource levels, MCP-OD personnel are unlikely to be fully deployable for short-notice missions, such as those in response to GRF orders. The impact of the entire MCP-OD not being available to deploy at the same time as their AC counterparts in the HQ, approximately 30 days later in the best case and possibly not for up to 270 days in the slowest case, depends on whether the division's command posts are augmented from other sources. Despite low MCP-OD fills, in none of the WFX AARs we observed was MCP-OD unavailability mentioned as a shortfall. Indeed, one division specifically mentioned exercising the MCP-OD as one of the WFX objectives, yet this was not addressed in the final AAR.
5. A definitive answer regarding the effect of MCP-ODs upon division HQ readiness is probably not feasible under current structure and regulations. Because the MCP-OD CUSR is submitted independently from the division HHBn CUSR, readiness reporting does not indicate *any* impact of the MCP-OD (positive or negative) on the division's level of readiness. Additionally, interviewees in each of the division WFX we visited reported their command posts were overstrength compared to MTOE authorizations or they had been augmented from units other than the MCP-OD. This made it problematic to discern the impact of the MCP-OD per se.
6. Doctrine and practice regarding division command post operations are evolving. The FARG II assumption of a home station command post and a deployed command post may be invalid. None of the division HQ in the WFX we observed operated in this manner.
7. The ability of MCP-ODs to conduct collective training and to integrate with the rest of a division HQ is affected not just by the number of training days available but also by synchronization challenges between AC and RC training management systems.

8. CA and MISO positions on the MCP-OD MTOE are currently shown as required but not authorized. We were unable to identify plans for filling those positions that would enable collective training prior to notification of sourcing.
9. Requirements to accomplish in-garrison tasks were inadequately considered in the FARG II design.

Recommendations

Our recommendations to the Army fall into three main categories: division structure, doctrine and guidance, and training and resourcing of the MCP-ODs.

Division Structure and Manning

1. The missions given to division HQs and the conditions under which they are performed are so varied that a study like this cannot recommend changes in the fundamental structure. However, there are three ways the process of documenting the structure might be changed to better achieve the operational objectives. The Army may want to consider two different division designs: one fully manned by the AC, focused on short-notice deployments across the spectrum of conflict and without a MCP-OD; and one that accepts the risks of the FARG II design (as mitigated by the MCP-OD).
2. The Army should also reconsider creating division HQ as true MCUs and integrating the MCP-ODs accordingly, versus the current designation as partner units. This would help address the current inconsistency in how the readiness of the overall HQ is assessed and reported.
3. TRADOC should document MCP-OD CA and MISO requirements on one USAR unit TOE and MTOE. Pending a more comprehensive redesign of CA and MISO TOEs, the easiest near-term solution might be to follow current practice and document those positions in a distinct cell within USACAPOC.
4. FORSCOM and the ARNG should examine split stationing options for MCP-ODs that will better align soldiers for both career progression and for appropriate training opportunities.
5. If, after several years, the MCP-ODs have consistent difficulties filling some positions due to the lack of sustainable personnel pipelines in their states, the MCP-OD MTOE could be converted to a true MCU and USAR soldiers assigned to fill the positions based on that component's core competencies.
6. Additionally, each division should develop a command post contingency staffing plan for filling critical position shortfalls if they are not sufficiently mitigated by their partnered MCP-ODs.

Doctrine and Guidance

At the time we were conducting the field research for this report, there remained numerous areas where "the field" seemed confused about the redesign effort and other places where doctrine writ large had not caught up with the changes at the HQ.

7. FORSCOM should consider promulgating an information paper or other communication on Army decisionmaking and division HQ force structure trade-offs. TRADOC should consider including division HQ design in the curriculum for intermediate-level

professional military education courses. The division HQ TOE (and MTOE) narratives should more explicitly lay out the FARG II design risk and the relationship between the division HQ and the MCP-OD.

8. FORSCOM should consider forming a study team to consider how Objective T readiness reporting might be better applied to MCP-OD CUSR reporting than the current process.
9. The Combined Arms Doctrine Directorate, U.S. Army CAC should incorporate MCP-OD considerations in the forthcoming revision of ATP 6-0.5, *Command Post Organization and Operations.*
10. FORSCOM should consider forming a study group to assess garrison support requirements for division HQ in the wake of FARG II reductions. For example, can additional tasks be contracted or assigned to echelons below division? Should some of these requirements be the responsibility of installation management command versus being tasked to divisions?

Main Command Post–Operational Detachments Training and Resourcing

A third set of recommendations focuses on what can be done within the current structure to maximize the readiness of the total division HQ.

11. The Army should consider designating one or more MCP-ODs as focused readiness units and resourcing them to enable deployment within 60 days of notification.
12. AC divisions with MCP-ODs should collaborate closely with MCP-OD commanders, the appropriate state Joint Forces HQ, and the USAR command as appropriate to (a) plan, execute, and validate training as required to meet deployment requirements within 270 days of notification of sourcing; (b) identify and document shortfalls given current RC resourcing levels; and (c) plan for postmobilization training requirements to close the identified shortfalls.
13. Division chiefs of staff and MCP-OD commanders should collaborate closely to synchronize AC and RC training management cycles to optimize MCP-OD readiness and integration into the division HQ. One approach to maximize cross-component training would be to have some AC soldiers in selected division command post sections periodically train on weekends to coincide with MCP-OD IDT and be given a four-day training holiday in exchange.

Implementing these recommendations, in part or in whole, will help the Army ensure division HQ will be as ready as possible to deploy under a wide range of mission parameters. Because the impetus for the FARG II changes came from outside the doctrine community and largely reflected an exogenous requirement to reduce authorizations, we did not assess or suggest any changes to the overall size of the HQ or its functional composition. As Chapter Two described, however, structure is rarely static, and at some point the Army will reconsider the decision to accept risk in the division HQ design. Until then, these types of improvements in how the MCP-OD and the division staff operate, across DOTMLPF-P, will provide additional mitigation of the risk the Army has so far accepted.

APPENDIX A

Aligning Active Component and Reserve Component Training Management

Numerous interviewees stated that the challenges posed by the MCP-OD concept include the differences between AC and RC training management cycles. A few soldiers reported that some AC leaders were insufficiently aware of these differences, thus resulting in a negative impact on MCP-OD training opportunities. In this appendix, we provide an overview of those differences.

DoD and Army policies require the Army to organize, man, train, and equip their active and reserve components as an integrated operational force.[1] Chapter Four discusses training within a DOTMLPF-P framework and notes that synchronization of AC and RC training cycles was a frequently reported issue. Both AC and RC soldiers are trained for their unique assignments and required readiness levels, but there are distinct differences between AC and RC training that create challenges for planners. AC planners working with MCP-ODs must consider these key differences, as well as valuable policy documents, to help best align AC and RC training management.

Two key policy documents for AC planners are DoDI 1215.06 (Uniform Reserve, Training, and Retirement Categories for the Reserve Component) and AR 350-1 (Army Training and Leader Development). It is important to note that these policy documents were last issued in 2015 and 2014, respectively. The findings of the National Commission on the Future of the Army were published on January 28, 2016. Ongoing implementation of the Army's Total Force Policy requires revision and consolidation of AR 350 series publications. Both implementation of National Commission on the Future of the Army findings and the ATFP may affect these and other training publications.

One key distinction affecting training is the general organization of the RC. RC soldiers are placed in one of three reserve component categories: the ready reserve, the standby reserve, and the retired reserve. Unlike AC soldiers who have one duty status (active duty), RC soldiers can serve in and switch between a myriad of statuses governed by a variety of laws and policies. In many cases, an RC soldier's duty status dictates the minimum training requirement, any restrictions on that training, and incidental considerations, such as pay caps and travel reimbursements. Figures A.1 and Table A.1 show the various RC duty statuses, including their governing authorities and general purposes.

Training for RC soldiers is accomplished during IDT, unit training assembly (UTA), multiple unit training assembly, drills, or AT.[2] Except active Guard/Reserves (AGRs), IDT requires

---

[1] U.S. Department of Defense, "Managing the Reserve Components as an Operational Force," directive no. 1200.17, October 29, 2008.

[2] U.S. Army War College, *How the Army Runs 2015–2016: A Senior Leader Reference Handbook*, Carlisle, Pa.: U.S. Army War College, 2015, p. 6-7.

Figure A.1
Reserve Component Duty Status Types

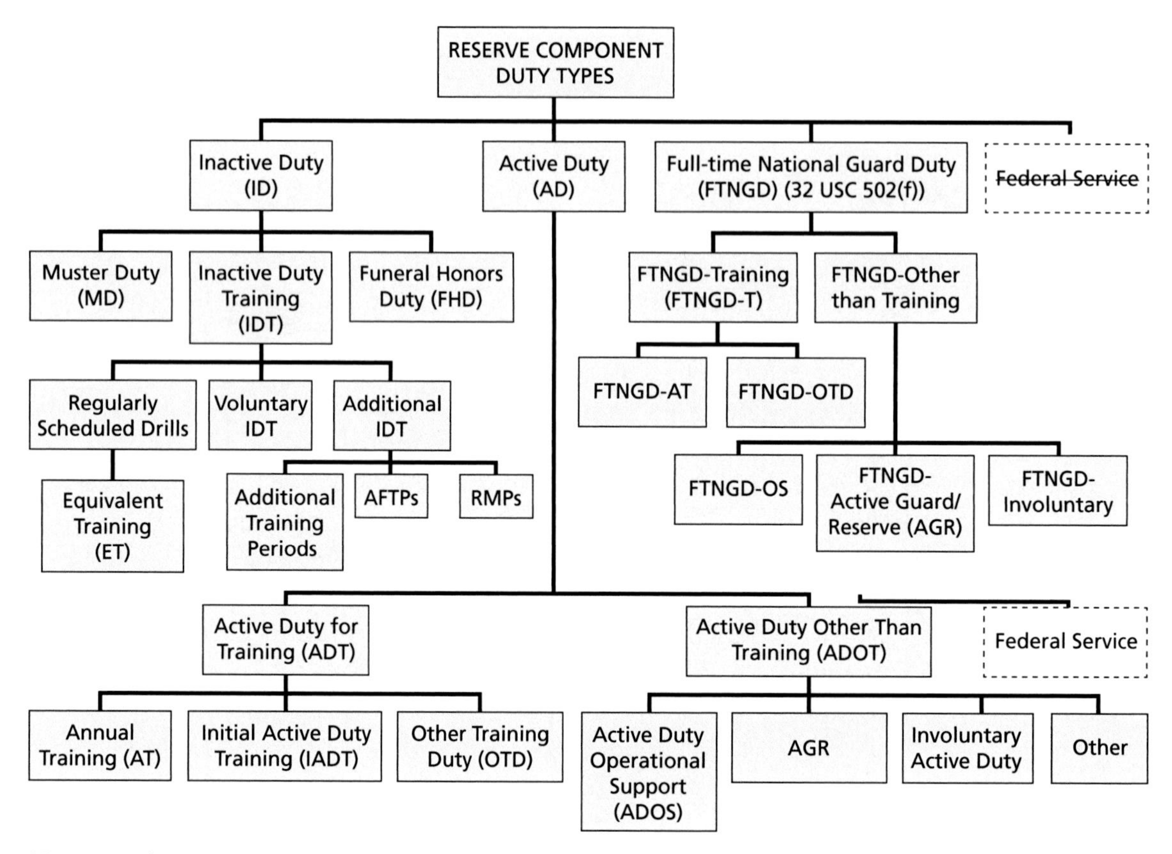

SOURCE: Enclosure 3, DoDI 1215.06, "Uniform Reserve, Training, and Retirement Categories for the Reserve Components."

participation in 48 scheduled drills or training periods each year.[3] RC individual training and weapons qualifications typically occur during IDT.[4] IDT periods are prescheduled and must last at least 4 hours. Additional IDT periods, such as additional training periods, are governed by military department policy, and they are capped at 36 each fiscal year. Although voluntary IDT is available, it is performed in a nonpay status.

USAR units are required to perform at least 14 days (exclusive of travel time) of AT each year, and it is geared primarily toward collective premobilization tasks. National Guard units are required by statute to perform AT for at least 15 days each year (inclusive of travel). AT periods for individual mobilization augmentees (another part of the selected reserve, separate from the USAR and ARNG unit structure) are typically funded for up to 12 days (inclusive of travel time), although this training can occur on any day of the week. AT periods are capped at 29 days per fiscal year, and they typically occur during one consecutive period. Split tours of AT can occur if required for selected individuals or units to meet their training missions or enhance DoD mission support, but split tours must first be authorized.

---

[3] DoDI 1215.06, Enclosure 7, para. 3(a)(1)(a).

[4] U.S. Army War College, 2015, p. 6-7.

Table A.1
Reserve Component Utilization Authorities

Legal Authority	Purpose of Duty	Applies to	Type of Duty	
Training				
10 USC 10147	AT drill requirement	Reserve only	AD/IDT	Involuntary
10 USC 12301(b)	AT	Reserve and National Guard	AD	Involuntary
10 USC 12301(d)	Additional/other training duty	Reserve and National Guard	AD	Voluntary
32 USC 502(a)	AT/drill requirement	National Guard only	FTNGD/ IDT	Involuntary
32 USC 502(f)(1)(A)	AT duty	National Guard only	FTNGD	Involuntary
32 USC 502(f)(1)(B)	Additional/other training duty	National Guard only	FTNGD	Voluntary
Support				
10 USC 12301(d)	AGR duty/operational support/additional duty	Reserve and National Guard	AD	Voluntary
10 USC 12304b	Preplanned/preprogrammed CCDR support	Reserve and National Guard	AD	Involuntary
32 USC 502(f)(1)(B)	AGR duty/operational support/additional duty	National Guard only	FTNGD	Voluntary
32 USC 502(f)(1)(A)	Other duty	National Guard only	FTNGD	Involuntary
Mobilization				
10 USC 12301(a)	Full mobilization	Reserve and National Guard	AD	Involuntary
10 USC 12302	Partial mobilization	Reserve and National Guard	AD	Involuntary
10 USC 12304	PRC	Reserve and National Guard	AD	Involuntary
10 USC 12304a	Emergencies and natural disasters	Reserve only	AD	Involuntary
14 USC 712	Emergencies and natural disasters	USCGR only	AD	Involuntary
Other				
10 USC 12503	Funeral honors	Reserve and National Guard	ID	Voluntary
32 USC 115	Funeral honors	National Guard only	ID	Voluntary
10 USC 12319	Muster duty	Reserve and National Guard	ID	Involuntary
10 USC 12301(h)	Medical care	Reserve and National Guard	AD	Voluntary

Table A.1—Continued

Legal Authority	Purpose of Duty	Applies to	Type of Duty	
10 USC 12322	Medical evaluation and treatment	Reserve and National Guard	AD	Voluntary
10 USC 12323	Pending LOD for response to sexual assault	Reserve and National Guard	AD	Voluntary
10 USC 688	Retiree recall	Reserve and National Guard	AD	Involuntary
10 USC 802(d)	Disciplinary	Reserve and National Guard	AD	Involuntary
10 USC 14108	Unsat participation (up to 45 days)	Reserve and National Guard	AD	Involuntary
10 USC 12301(g)	Captive status	Reserve and National Guard	AD	Involuntary
10 USC 12303	Unsat participation (up to 24 months)	Reserve and National Guard	AD	Involuntary
10 USC 12402	Duty at National Guard bureau	National Guard only	AD	Voluntary
10 USC 331	Insurrection	National Guard only	AD	Involuntary
10 USC 332	Insurrection	National Guard only	AD	Involuntary
10 USC 12406	Insurrection	National Guard only	AD	Involuntary

SOURCE: Appendix to Enclosure 4, DoDI 1215.06, "Uniform Reserve, Training, and Retirement Categories for the Reserve Components."

KEY: AD = active duty; CCDR = combatant commander; LOD = line of duty; PRC = Presidential reserve call-up.

Another distinction between AC and RC training involves timing of the training. Because most RC soldiers have civilian employers, training is scheduled to minimize the disruption to that civilian employment. Typically, UTAs are consolidated, so four UTAs are accomplished during a single weekend.[5] Timing considerations also apply to when an RC soldier receives certain types of training. RC soldiers receive their unit assignment prior to completion of MOS qualification training. The Army War College's handbook, *How the Army Runs*, notes that for RC soldiers, "qualification training, sustainment training, additional duty training, and professional development education are often conducted in lieu of scheduled UTAs and AT, and in some cases require more than a year to complete."[6] While completing this training, they are not available to participate in collective training. Because most RC soldiers have civilian employment, they also have shorter time frames in which to complete training requirements to meet the same standards as their AC counterparts.

[5] U.S. Army War College, 2015, p. 6-7.

[6] U.S. Army War College, 2015, p. 6-7.

Training cycles not only differ for the individual soldiers but also among the components. Field Manual (FM) 7-0 (*Train to Win in a Complex World*) describes some planning considerations for the differing short-, mid-, and long-range AC and RC training horizons. It notes that the primary areas where these differences occur are in planning horizons, resource coordination, cyclical training briefings, and utilization of the T-week concept.[7]

Compared to the AC, the RC has fewer training days, geographically dispersed soldiers with competing civilian commitments, and different resource pools, all of which necessitate longer planning and notice periods than those within the AC. The AC typically plans its training on a quarterly basis, but the RC operates on a yearly training calendar. Tables A.2 and A.3 show the different long-range planning horizons for the AC and RC.

To ensure RC participation in training events, AC and RC planners must begin planning much earlier and avoid last-minute changes or updates.

Planners must also consider the time lines and restrictions on pre- and postmobilization training for RC soldiers. Defense Secretary Gates's 2007 memo on utilization of the reserve components both limited involuntary mobilizations to a maximum of one year and set a targeted MOB to DWELL ratio of one to five.[8] This policy change and the Army's Execution Order 150-08 in 2008 rebalanced the premobilization and postmobilization training that RC soldiers receive. Premobilization training was focused on completing individual training and readiness activities and conducting collective training to the extent possible.[9] A 2015 RAND report found that despite resources dedicated to increased premobilization training after 2008, RC units found it difficult to complete all individual training

Table A.2
Regular Army Long-Range Planning by Echelon

Echelon	Publishes CTG with Calendar NLT[a]	Planning horizon
Corps	12 months prior to training start	2 years
Division	10 months prior to training start	2 years
Installation	10 months prior to training start (calendar only)	1 year
Brigade	8 months prior to training start	1 year
Battalion[b]	6 months prior to training start	1 year

SOURCE: Headquarters, Department of the Army, October 2016.

[a] Publication dates also apply to similar command-level TDA organizations or activities. For example, a TRADOC COE normally commanded by a major general follows the same planning cycle as a division commander.

[b] Companies develop and publish their own UTP. The battalion commander, in collaboration with subordinate company commanders and the battalion staff may develop a consolidated battalion UTP.

KEY: COE = center of excellence; CTG = command training guidance; NLT = no later than; UTP = unit training plan.

---

[7] Headquarters, Department of the Army, FM 7-0, October 2016, p. 1–18.

[8] Secretary of Defense, January 19, 2007.

[9] Ellen M. Pint, Matthew W. Lewis et al., *Active Component Responsibility in Reserve Component Pre- and Postmobilization Training*, Santa Monica, Calif.: RAND Corporation, RR-738-A, 2015, p. 47.

Table A.3
Reserve Component Long-Range Planning by Echelon

Echelon	Publishes CTG with Calendar NLT[a]	Planning Horizon
Flag officer CMD, separate brigade, regiment or group	18 months prior to training start	5 years
Brigade or separate battalion	10 months prior to training start	5 years
Battalion[b]	6 months prior to training start	2–3 years

SOURCE: Headquarters, Department of the Army, October 5, 2016.

[a] These actions also apply to similar command-level TDA organizations or activities. For example, a regional support command, commanded by a major general follows the same planning cycle as a division commander.

[b] Companies develop and publish their own UTP. The battalion commander, in collaboration with subordinate company or troop commanders, and the battalion staff may develop a consolidated battalion UTP.

requirements prior to mobilization.[10] Reasons for the individual training shortfalls include the availability of equipment to RC soldiers for qualification and inability to attend all AT due to civilian job commitments.[11] Although these training shortfalls were made up post-mobilization without delay of arrival in theater, they highlight the necessity of close coordination between all components to achieve readiness for all soldiers in an MCP-OD.

[10] Pint, Lewis et al., 2015, p. 54.

[11] Pint, Lewis et al., 2015, p. 55.

Division Headquarters Fill Rate Tables

When beginning this research, we noted that the question of how to fill HQ did not begin with the FARG II design. We therefore thought it was important to look at the recent history of AC staffing, to consider its effect on the HQ's readiness. From the available data, we see that, in general, the Army is less likely to take risks in manning the Fires and Intel cells, while accepting risk in some low-density MOSs. As a result, when if a HQ is called to deploy rapidly and at a high percentage of its authorized strength, the same system used to address MCP-OD shortages and delays will probably be used to fill AC vacancies.

This appendix presents personnel fill rates (percentage of authorized personnel who are on hand) by MOS and AOC for five division HQ, which we have numbered 1 to 5. Ideally, if a division HQ has a shortfall in any MOSs/AOCs, the MCP-OD associated with that division HQ will have the specialties needed to address those gaps. Earlier in this report, we looked at personnel fill rates of MCP-ODs 1, 2, and 3. Here we look at personnel fill rates of division HQ 1 to 5.

When we looked at MCP-OD fill rates, we focused on a snapshot of MCP-OD fill rates at the end of FY 2016. (Snapshots for three MCP-ODs from September 2016 were used due to Defense Manpower Data Center data constraints—and because only three MCP-ODs had most or all of their personnel at that point.) In the case of division HQ, however, we could use TAPDB data to examine fill rates over time, typically from FY 2009 to FY 2016 (with the period depending on the availability of historical data).

Table B.1 shows the MOSs/AOCs that had personnel challenges in Division HQ 1. The first column lists the MOS/AOC. The second column shows the average MTOE authorizations for that MOS/AOC from FY 2009 to FY 2016. The third column shows that their average fill was below 90 percent between FY 2009 and FY 2016. The fourth shows how the average fill changed during the unit's deployment in FY 2015. The fifth column shows the most recent fill rate in the data (an indicator of whether fill is still an issue). The sixth and seventh columns show the coefficient of variation (standard deviation divided by mean) of fill; these figures indicate how much the fill varied over time (across months) during FY 2015 and FY 2016.

Two specialties with *particularly* low fill rates (equal to or below 50%) overall and during FY 2016 were 49A and 50A, with average personnel fill rates of 38 percent and 49 percent, respectively. However, other MOSs/AOCs also had shortfalls (e.g., 255N [network management technician], 29A [electronic warfare officer], and 38A [CA]). A subset of MOSs/AOCs had higher fill rates during deployment, compared to their average fill rates overall: 255N (network management technician), 74A (chemical, biological, radiological, and nuclear), and 89E

Table B.1
MOS/AOC Fill Rates—Division HQ 1

MOS/AOC—DIV HQ #1	Average Number Authorized FY 2009–FY 2016	Mean Fill Rate FY 2009–FY 2016	Mean Fill Rate During FY 2015 (Deployment)	Mean Fill Rate in FY 2016	CV in FY 2015	CV in FY 2016
255N: Network management technician	3	0.54	0.81	0.47	0.28	0.56
29A: Electronic warfare officer	2	0.55	0.58	0.46	0.88	0.87
30A: Information operations officer	4	0.84	0.38	0.61	0.35	0.21
35T: Military intelligence systems maintainer/integ	3	0.74	0.19	0.69	0.49	0.32
38A: Civil affairs	3	0.64	0.56	0.83	0.30	0.30
40A: Space operations	2	0.72	0.67	0.11	0.67	1.48
49A: Operations research/systems analysis	2	0.38	0.29	0.17	0.88	1.48
50A: Force development	1	0.49	0.00	0.33	—	1.48
70E: Patient administration	1	0.67	0.58	0.17	0.88	2.34
72D: Environmental science and engineering	1	0.67	0.58	0.42	0.88	1.24
74A: Chemical, biological, radiological, and nuclear	4	0.88	1.06	0.67	0.20	0.21
89E: Explosive ordnance disposal	1	0.62	1.00	0.83	0.00	0.47
91C: Utilities equipment repairer	4	0.54	0.31	0.67	0.36	0.00

SOURCE: RAND Arroyo Center analysis of FMSWeb data.

(explosive ordnance disposal). Specialties with coefficients of variation greater than 1 during FY 2015 and/or FY 2016—for example, 49A (operations research), 70E (patient administration), and 72D (environmental science and engineering)—had particularly variable fill rates during those years.[1]

Table B.2 shows the MOSs/AOCs that had personnel challenges in Division HQ 2.[2] Some of the same specialties that had low fill rates in Division HQ 1 also had low fill rates in Division HQ 2 (e.g., 255A, 29A, 40A, 49A, and 72D). Specialties in Division HQ 1 with particularly low average fill rates (equal to or below 50%) over time were 150A (air traffic and space management technician), 153A (rotary wing aviator), 255A, 29A, and 57A (simula-

[1] A coefficient of variation greater than 1 means the standard deviation of fill exceeded the average fill during the fiscal year. For example, the TAPDB data indicated that during FY 2015, the patient administration position (70E) was filled by a health services officer (MOS 67A) from October 2014 until April 2015—but not from May 2015 through September 2015. Additionally, in FY 2016 the slot was filled in May and June of 2016—but not in other months.

[2] In Division HQ 2, the fill rates for officer AOCs and enlisted MOSs are based on TAPDB data from FY 2009 through FY 2016. However, for this division, we only had warrant officer TAPDB data from FY 2011 to FY 2013; thus, the fill percentages for 150A, 153A, 255A, and 255N are based on more limited data.

Table B.2
MOS/AOC Fill Rates—Division HQ 2

MOS/AOC—DIV HQ #2	Average Number Authorized FY 2009– FY 2016	Mean Fill Rate FY 2009– FY 2016	Mean Fill Rate During FY 2015 (Deployment)	Mean Fill Rate in FY 2016	CV in FY 2015	CV in FY 2016
150A: Air traffic and air space management technician	2	0.46	0.58	0.59	0.00	0.33
153A: Rotary wing aviator (aircraft nonspecific)	2	0.32	0.21	0.18	0.73	1.24
255A: Information services technician	5	0.36	0.40	0.40	0.68	0.43
255N: Network management technician	3	0.54	0.67	0.67	0.84	0.48
29A: Electronic warfare officer	2	0.28	0.08	0.58	0.45	0.72
29E: Electronic warfare specialist	1	0.78	0.67	0.75	0.69	0.60
40A: Space operations	2	0.82	1.17	0.36	0.30	0.48
49A: Operations research/ systems analysis	2	0.74	0.63	0.58	0.32	0.33
57A: Simulations operations officer	2	0.51	0.50	0.47	0.86	0.56
72D: Environmental science and engineering	1	0.86	1.25	0.33	0.60	1.48
92M: Mortuary affairs specialist	1	0.85	1.00	0.17	0.74	2.34

SOURCE: RAND Arroyo Center analysis of FMSWeb data.

tion operations). Specialties with high coefficients of variation (particularly variable fill rates) during FY 2016 were 153A, 72D, and 92M (mortuary affairs specialist).

Table B.3 shows the MOSs/AOCs that had personnel challenges in Division HQ 3. Some of the previously mentioned shortfalls (associated with Division HQ 1 and 2) also appear in Division HQ 3: 153A, 255A, 29A, 38A, 40A, 49A, 50A, and 72D. Specialties with particularly low fill rates were 153A, 29A, 38A, and 89D. Several medical specialties and munitions-related specialties had particularly variable fill rates (high coefficients of variation) in FY 2016. Although 89B (ammunition specialist) had the highest coefficient of variation, that was because the position was unfilled in all but one month of FY 2016, September 2016, so it had a low mean fill during FY 2016 (0.08), which made the coefficient of variation (standard deviation divided by mean) particularly high that year.

Table B.4 shows the MOSs/AOCs that had personnel challenges in Division HQ 4. (We only examined data from FY 2008 until FY 2014 for this division because this particular division transitioned to an MCU and changed UICs in 2015.) Division HQ 4 was deployed from February 2013 until January 2014. (Column 4 shows fill during FY 2013, which overlapped much of that period.)

On average, from FY 2008 until FY 2014, the specialties with particularly low fill in Division HQ 4 included 29A (electronic warfare officer) and 38 B (CA specialist). In the most recent

Table B.3
MOS/AOC Fill Rates—Division HQ 3

MOS/AOC—DIV HQ #3	Average Number Authorized FY 2009–FY 2016	Mean Fill Rate FY 2009–FY 2016	Mean Fill Rate During Deployment (mid 2012–mid 2013)	Mean Fill Rate in FY 2016	CV in FY 2015	CV in FY 2016
120A: Construction engineering technician	1	0.75	0.83	0.33	0.00	1.48
153A: Rotary wing aviator (aircraft nonspecific)	2	0.44	0.04	0.58	0.21	0.88
15B: Aircraft powerplant repairer	7	0.65	1.39	0.72	0.15	0.14
255A: Information services technician	5	0.52	0.12	0.38	0.08	0.35
255N: Network management technician	3	0.74	0.81	0.44	0.25	0.67
25S: Satellite communication systems operator—maintenance	21	0.80	0.96	0.76	0.06	0.13
29A: Electronic warfare officer	2	0.38	0.50	0.38	0.47	0.60
30A: Information operations officer	4	0.68	1.00	0.69	0.42	0.14
35L: Counter intelligence agent	4	0.74	0.73	0.88	0.13	0.15
38A: Civil affairs	3	0.35	0.33	0.63	0.00	0.77
40A: Space operations	2	0.57	1.00	0.42	0.60	0.36
46A: Public affairs, general	2	0.69	1.13	0.67	0.00	0.49
49A: Operations research/systems analysis	2	0.52	0.63	0.54	0.21	0.27
50A: Force development	1	0.82	1.00	0.67	1.24	0.74
53A: Information systems management	4	0.64	0.75	0.64	0.30	0.41
57A: Simulations operations officer	2	0.72	1.00	0.86	0.14	0.20
60W: Psychiatrist	1	0.71	1.00	0.25	0.00	1.81
67J: Aeromedical evacuation	1	0.63	1.00	0.33	1.60	1.48
70E: Patient administration	1	0.68	1.00	0.50	0.35	1.04
70H: Health services plans, operations, intelligence	2	0.70	0.67	0.75	0.00	0.53
70K: Health services materiel	1	0.76	1.00	0.58	0.00	0.88
72D: Environmental science and engineering	1	0.68	1.00	0.58	0.00	0.88
89B: Ammunition specialist	1	0.62	1.00	0.08	0.62	3.46
89D: Explosive ordnance disposal specialist	1	0.40	0.58	0.42	0.00	1.24
89E: Explosive ordnance disposal	1	0.51	1.17	0.42	0.00	1.24

SOURCE: RAND Arroyo Center analysis of FMSWeb data.

Table B.4
MOS/AOC Fill Rates—Division HQ 4

MOS/AOC—DIV HQ #4	Average Number Authorized FY 2008–FY 2014	Mean Fill Rate FY 2008–FY 2014	Mean Fill Rate During FY 2013 (Deployment)	Mean Fill Rate in FY 2014	CV in FY 2013	CV in FY 2014
24A: Telecommunications system engineer	2	0.88	0.92	0.63	0.21	0.69
255A: Information services technician	5	0.62	0.80	0.50	0.00	0.55
29A: Electronic warfare officer	2	0.19	0.33	0.67	2.34	0.67
29E: Electronic warfare specialist	1	0.77	1.00	0.67	0.00	0.74
352N: Signals intelligence analysis technician	2	0.78	1.00	0.83	0.00	0.30
37A: Psychological operations	2	0.73	0.58	0.50	0.33	0.43
38A: Civil affairs	3	0.52	0.33	0.47	0.36	0.36
38B: Civil affairs specialist	3	0.46	0.58	0.69	0.80	0.14
49A: Operations research/systems analysis	2	0.81	0.63	0.38	0.36	0.83
57A: Simulations operations officer	2	0.71	0.50	0.50	0.00	0.74
59A: Strategist	2	0.70	0.42	0.63	0.21	0.47
70E: Patient administration	1	0.75	0.58	0.50	0.88	1.35
70H: Health services plans, operations, intelligence	3	0.88	0.63	0.75	0.00	0.60
70K: Health services materiel	1	0.88	0.75	0.58	0.00	0.88

SOURCE: RAND Arroyo Center analysis of FMSWeb data.

of those years (FY 2014), 49A (operations research) and 38A (CA) also had particularly low fill. When Division HQ 4 was deployed, specialties related to electronic warfare and communication tended to have greater fill than at other times. Specialty 70E (patient administration) tended to have more fluctuation in fill rate (higher coefficient of variation) than other MOSs.

Table B.5 shows the MOSs/AOCs that had personnel challenges in Division HQ 5. Division HQ 5 was deployed during FY 2010. Some MOSs that had low fill rates in later years were not on the FY 2009 or FY 2010 MTOE; thus, we did not show a fill rate for them during the FY 2010 deployment.

The preceding data show MOS fill is not just an RC issue; it is also an AC issue. These figures show that the Army seems to have habitually been willing to accept risk in manning of division HQ before deployment. We also see that, in general, the Fires and Intel cells are areas where the Army is less likely to take risk with AC manning. Furthermore, most issues seem to be in low-density MOSs. Regardless of the presence or absence of a ready or not-ready MCP-OD, Army divisions should expect to be alerted and potentially deploy with gaps in personnel and skills (capability and capacity), and the system that is in place to mitigate or address these challenges may also be of use in mitigating/addressing MCP-OD shortages/delays.

Table B.5
MOS/AOC Fill Rates—Division HQ 5

MOS/AOC—DIV HQ #5	Average Number Authorized FY 2009–FY 2016	Mean Fill Rate FY 2009–FY 2016	Mean Fill Rate During FY 2010 (Deployment)	Mean Fill Rate in FY 2016	CV in FY 2015	CV in FY 2016
153A: Rotary wing aviator	2	0.42	— (not on FY 2010 MTOE)	0.04	1.04	3.46
255A: Information services technician	5	0.65	—	0.65	0.16	0.20
255N: Network management technician	3	0.60	—	0.33	0.13	0.00
255S: Information protection technician	1	0.17	—	0.33	—	1.48
25L: Cable systems installer–maintainer	16	0.86	—	0.86	0.05	0.05
36B: Financial management technician	3	0.86	0.46	0.86	0.00	0.20
37A: Psychological operations	2	0.82	1.83	0.83	0.16	0.30
38A: Civil affairs	3	0.70	0.79	0.75	0.14	0.20
38B: Civil affairs specialist	3	0.05	—	0.31	—	0.73
59A: Strategist	2	0.41	—	0.79	0.37	0.33
67J: Aeromedical evaluation	1	0.55	0.83	0.08	—	3.46
89D: Explosive ordnance disposal specialist	1	0.50	—	0.17	0.00	2.34

SOURCE: RAND Arroyo Center analysis of FMSWeb data.

A Short History of U.S. Division Headquarters Deployments Since the End of the Cold War

Chapter Two summarized the evolving nature of Army division HQ deployments. In this appendix, the authors provide details from some of these deployments to illustrate and expand upon the generalizations provided earlier.

Operations Restore Hope and Continue Hope, Somalia, 10th Mountain Division, December 1992–March 1994

In 1992 U.S. forces, in conjunction with other national forces and the UN, entered Somalia to establish security for humanitarian relief.[1] The U.S. mission was code-named Restore Hope and transitioned to Continue Hope in May 1993.[2] The 10th Mountain provided the forces for Task Force Mountain, consisting of two infantry battalions, an aviation brigade, the division artillery, and support assets. At its peak strength, Task Force Mountain had approximately 10,000 soldiers.[3] The 10th Mountain was also designated ARFOR HQ for the mission. ARFOR provided security in the areas around Kismaayo, Merca, Baledogle, and Baidoa.[4] Army units were deployed and were conducting their security missions by early January 1993.[5] The 1st Brigade, 10th Mountain, went on to assume the important role of Quick Reaction Force in May 1993. Following the decision to withdraw U.S. forces in late 1993, 10th Mountain provided the core staff of JTF Somalia. It was activated on October 14, 1993, and became fully operational six days later.[6] Due to constraints on size imposed by U.S. Central Command, it was limited to 60 personnel.[7] While the intention to withdraw had been announced, the JTF continued to conduct operations, including two humanitarian relief missions and the orderly handover to UN forces. The full tactical withdrawal of U.S. forces was completed by March 25, 1994.

---

[1] U.S. Department of Defense, *United States Forces, Somalia After Action Report and Historical Overview: The United States Army in Somalia, 1992–1994*, Washington, D.C.: U.S. Army Center of Military History, 2003, p. 5.

[2] For the official history of the operation, see Poole, 2005; U.S. Department of Defense, 2003.

[3] U.S. Department of Defense, 2003, p. 6.

[4] U.S. Department of Defense, 2003, pp. 6–7.

[5] U.S. Department of Defense, 2003, p. 6.

[6] U.S. Department of Defense, 2003, pp. 141–142.

[7] U.S. Department of Defense, 2003, p. 224.

The role of the division HQ is important in the Somalia operations because it was the first time a division had taken on this "operational"-level role.[8] The formal AAR noted concerning the JTF that while there was a good balance of tactical and technical expertise in terms of division actions, there was very limited knowledge of the difficulties of joint operations.[9] A 2000 RAND study found that serving as the senior "tactical" HQ in Somali in 1992 "'stretched' the 10th Mountain's command capabilities in four ways, exposing shortfalls in both training and technology normally available at the division level."[10] Four particular stress points were

1. greater span of control than normal: "It took several corps- and theater-level signal companies simply to tie the overall unit together"
2. wide geographical dispersion: units spread over a 100-kilometer "front," beyond the line-of-sight communications systems then standard in the Army
3. unfamiliar tasks: as ARFOR HQ to a JTF commanded by a U.S. Marine Corps general, the division took on tasks "that in war would normally be handled by higher Army echelons"
4. sizable political-military challenges, specifically the need to support numerous allied forces attached to the division and to coordinate with nongovernmental organizations and nondefense U.S. government agencies.

Figure C.1 depicts the 10th Mountain forces deployed to Somalia.

Figure C.1
U.S. Army Forces in Somalia

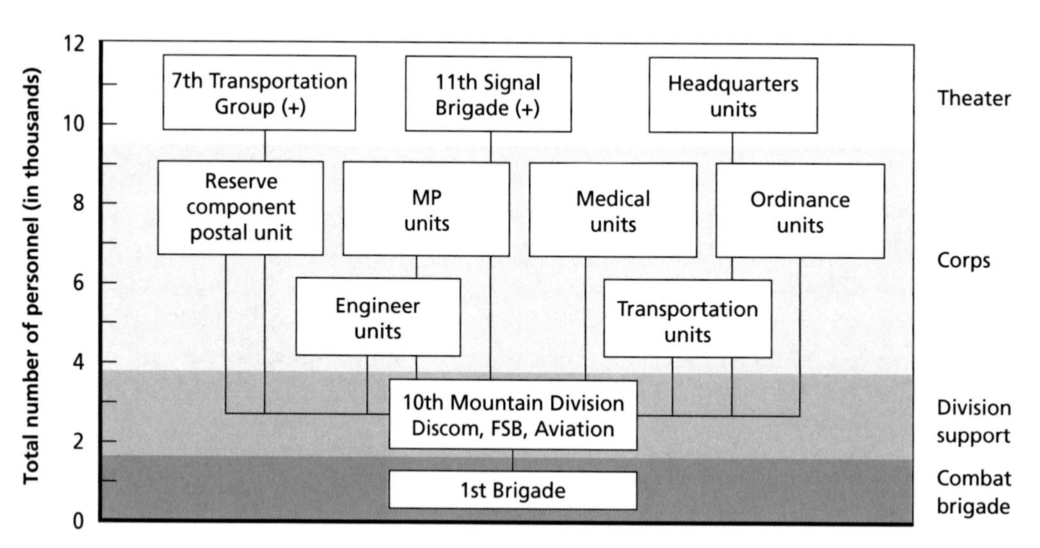

SOURCE: McNaugher, Johnson, and Sollinger, 2000, p. 2.
KEY: FSB = Forward Support Battalion.

---

[8] Jacobi, 2004, p. 6.

[9] U.S. Department of Defense, 2003, pp. 224.

[10] Thomas L. McNaugher, David E. Johnson, and Jerry M. Sollinger, *Agility by a Different Measure: Creating a More Flexible U.S. Army*, Santa Monica, Calif.: RAND Corporation, IP-195, 2000, p. 2.

Equally relevant to this study, McNaugher, Johnson, and Sollinger, went on to describe a fifth challenge:

> The division drew supporting units from across the United States. . . . These units had no opportunity to train together before arriving in Mogadishu. Operating procedures had to be developed on the fly, resulting in a certain amount of "friction" early in the operation.[11]

This challenge refers to subunits, not staff sections, and to a doctrinal context that did not prioritize modularity. However, they noted that the 10th Mountain in Haiti and the 1st Cavalry Division in Bosnia (1998–1999) required corps and theater army augmentation.[12]

Operation Uphold Democracy, Haiti, 10th Mountain Division and 25th Infantry Division, July 1994–March 1995

On July 29, 1994, the 10th Mountain's HQ was "read into" the existing planning for operations in Haiti. The U.S. military had been planning and preparing for a range of operational scenarios for some time in response to the 1991 coup d'état that overthrew President Jean-Bertrand Aristide. Late in the planning cycle, the 10th Mountain was designated JTF 190 and given responsibility for developing a plan for a permissive entry into Haiti, by air and sea, called OPLAN 2380.[13] This was in parallel to the efforts of XVIII Airborne Corps, which was developing a separate plan for an airborne assault, OPLAN 2370.[14]

The final division plan was published on September 1, and on September 9 the 10th Mountain and the XVIII Airborne Corps received the activation order. On September 16, the President notified the military that he had decided to implement the Haiti military operation. On September 18, the XVIII Airborne Corps received the approval of OPLAN 2370 and began deploying. However, a last-minute peace mission led by President Jimmy Carter was successful, and the operation was canceled with elements of the XVIII Corps in midair on their way to conduct the operation. Under the peace deal, the existing Haitian government would remain in power to facilitate a transition and democratic elections, and U. S. forces would enter and ensure the security of the transition.

On September 19, the 1st BCT of the 10th Mountain conducted an air assault on Port-au-Prince International Airport, and between September 20 and 28, follow-on elements of the 10th Mountain arrived in-country.[15] The division was tasked to control Port-au-Prince and Cap-Haïtien and conducted patrols in the surrounding areas. Concurrently U.S. Special Forces and other units, including U.S. Marines, conducted operations on the island. The 10th Mountain was designated the Joint Force Command (JFC). As the JFC, the 10th Mountain also served as the multinational force HQ and was responsible for the reception, staging, and integration and supervision of the many multinational units arriv-

---

[11] McNaugher, Johnson, and Sollinger, 2000, p. 2.

[12] McNaugher, Johnson, and Sollinger, 2000, p. 2.

[13] W. E. Kretchick, R. F. Baumann, and J. T. Fishel, *Invasion, Intervention, "Intervasion": A Concise History of the U.S. Army in Operation Uphold Democracy*, Ft. Leavenworth, Kan.: U.S. Army Command and General Staff College Press, 1998, p. 58.

[14] Kretchick, Baumann, and Fishel, 1998, p. 63.

[15] Kretchick, Baumann, and Fishel, 1998, p. 101.

ing in theater.[16] The 25th ID received an oral warning order in early November 1994 that it would replace 10th Mountain in Haiti and formal notification on December 4, 1994, that on December 26 it would deploy 3,500 soldiers to continue peace operations and transition the operation to the UN.[17] The operation transitioned to a UN-led mission as of March 31, 1995, with U.S. MG Joseph Kinzer assuming the dual role of U.S. force commander and UN force commander.[18]

It was 43 days from being stood up on July 29, 1994, to the activation order on September 9 and a further 10 days until 10th Mountain landed at Port-au-Prince. The role of JTF Operation Uphold Democracy was a significant expansion of a division HQ's responsibilities and something that 10th Mountain was not trained or doctrinally organized to do.[19] During the planning and preparation phase, 10th Mountain staff grew rapidly from 300 to 800.[20] In 1996, as a result of the Army experience on contingency operations, Army Doctrine, FM 71-100 Division Operations, was changed to include the role of ARFOR HQ for divisions. However, doctrine maintained that divisions would not normally be designated as JTFs.[21]

Operation Enduring Freedom, Afghanistan, March 2002–April 2005

Following the deployment and success of U.S. Special Forces in Afghanistan in 2001, the 10th Mountain, 82nd Airborne, and 101st Airborne were among the main divisions deployed to take command of U.S. forces in Afghanistan between 2002 and 2005. Initially the 10th Mountain deployed as Combined Forces Land Component Command (CFLCC) Forward/Task Force Mountain to Uzbekistan in December 2001 to take command of the buildup and ongoing operations.[22] In February 2002 the HQ moved to Bagram, Afghanistan. Both the span of control and the range units under CFLCC Forward/Task Force Mountain were unprecedented. They included conventional forces, special forces, and multinational forces, operating across Afghanistan. CFLCC Forward worked under the CFLCC Main, based in Kuwait, consisting of the Third Army HQ.[23] The official Army history of this period noted that the HQ deployed to Afghanistan as CFLCC were "essentially a stripped-down division HQ, the task force proved increasingly ill-suited to coordinate all conventional and unconventional operations taking place across Afghanistan."[24]

The capability and capacity of the division HQ elements in Afghanistan was to remain an issue through the period 2002–2005, due to the changing nature of the mission and the concurrent demands of the Iraq campaign.[25] During this period Iraq accounted for 90 percent

[16] Kretchick, Baumann, and Fishel, 1998, p. 105.

[17] Kretchick, Baumann, and Fishel, 1998, p. 135.

[18] Kretchick, Baumann, and Fishel, 1998, p. 136.

[19] Kretchick, Baumann, and Fishel, 1998, p. 63.

[20] Kretchick, Baumann, and Fishel, 1998, p. 100.

[21] Jacobi, 2004, p. 6.

[22] B. F. Neumann, L. Mundey, and J. Mikolashek, *The U.S. Army in Afghanistan Operation Enduring Freedom, March 2002–April 2005*, Washington, D.C.: Center for Military History, United States Army, 2013, p. 9.

[23] Neumann, Mundey, and Mikolashek, 2013, pp. 8–9.

[24] Neumann, Mundey, and Mikolashek, 2013, p. 11.

[25] Neumann, Mundey, and Mikolashek, 2013, p. 38.

of U.S. forces deployed in the U.S. Central Command region.[26] In response, the Army made several changes to the command structure. The first of these was to deploy XVIII Airborne Corps in April 2002 to take over as Combined JTF, to which the division HQ would be subordinate to throughout subsequent rotations.[27] In 2003 the staff of the 10th Mountain HQ was merged with the XVIII Airborne Corps command, creating Combined JTF-180 with a tactical HQ underneath, based on a division.[28] In late 2002, the 82nd Airborne arrived to take over from Task Force Mountain and provide the tactical HQ Combined JTF-82.[29] Beyond 2005 the mission in Afghanistan was to grow, especially after the drawdown in Iraq and the expansion of the NATO mission. From 2010 onward several division HQ rotated to Afghanistan to complete 12-month tours. Even after the end of Operation Enduring Freedom in 2015, the United States has maintained a division HQ as U.S. National Support Element and as HQ U.S. Forces-Afghanistan under Operation Freedom's Sentinel.

Operation Iraqi Freedom, 3rd Infantry Division "Task Force Marne," April 2007–June 2008

Of the many division deployments to Iraq, one of the most useful examples for this study is the deployment of the 3rd ID to Iraq in March 2007 for three reasons. First, it was a 15-month tour. Second, 3rd ID was one of the first units to undergo the transformation to a modular force system, with the addition of another brigade to the division and the transfer of division support elements to the brigades.[30] Third, the deployment coincided with the increase in violence, the U.S. surge, and the subsequent transition to reconciliation, one of the most challenging periods in Operation Iraqi Freedom.[31]

The deployment was announced in November 2006.[32] The 3rd ID would be the first unit to serve three tours in that country under Operation Iraqi Freedom. The division had previously helped take Baghdad in the initial invasion and had completed a rotation.[33] Initially the 3rd ID was due to deploy by June 2007, but in February the DoD announced that the division would deploy by March.[34] This reduced the divisions training time from six months to three weeks.[35] In addition, the division had been preparing to deploy to the north of Baghdad, but instead, in response to a change of campaign plan, the 3rd ID were tasked to deploy to the south of Baghdad and assume responsibility for a newly created command, Multi-National Division-Center. Multi-National Division-Center had a total area of more than 60,000 square

---

[26] Neumann, Mundey, and Mikolashek, 2013, p. 38.

[27] Neumann, Mundey, and Mikolashek, 2013, p. 13.

[28] Neumann, Mundey, and Mikolashek, 2013, p. 40.

[29] Neumann, Mundey, and Mikolashek, 2013, p. 16.

[30] D. Andrade, *Surging South of Baghdad: The 3D Infantry Division and Task Force Marne in Iraq, 2007–2008*, Washington, D.C.: U.S. Army Center of Military History, 2010, p. 4.

[31] For the official account of 3rd ID's deployment, see Andrade, 2010.

[32] Andrade, 2010, p. 384.

[33] Andrade, 2010, p. 5.

[34] U.S. Army Office of Public Affairs, "Headquarters 3rd Infantry Division—'Rock of the Marne'—Goes to Iraq in March," Army.mil, February 16, 2007.

[35] Andrade, 2010, p. 25.

kilometers and the deployment of 3rd ID, with its mix of conventional fighting power, doubled the size and capability of U.S. forces in the area.[36] The HQ of 3rd ID consisted of around 1,000 troops and commanded around 13,000 troops in theater, designated Task Force Marne.[37] About 60 percent of the soldiers in the division HQ had deployed to Iraq once, many twice.[38] The deployment is an example of a division and its HQ on continuous operations. The area of responsibility and forces assigned to it grew throughout the tour, and between June 2007 and June 2008 the division conducted 12 division-level operations.[39] The short-notice change of deployment time line and mission set a reminder that it is not just in contingency operations that a division must be ready to respond to last-minute changes. On July 5, 2008, 3rd ID returned home.

Operation United Assistance–Liberia, 101st Airborne Division, September 2014–February 2015

On September 16, 2014, in response to the Ebola virus disease (EVD) outbreak in West Africa, the President announced that the United States would deploy 3,000 troops to support the local and international efforts to combat the disease.[40] On September 26 the SECDEF approved the mission for the 101st Airborne to establish a JFC in Liberia. Up until the day before, the 101st Airborne was planning and conducting predeployment training in expectation of deploying to a "decisive action rotation" in late 2015.[41] Prior to the deployment of the 101st Airborne, preparatory work for JFC Operation United Assistance (UA) was carried out by personnel from U.S. Army Africa, the ASCC for U.S Africa Command. Within days, personnel from the 101st Airborne were deployed, and on October 24, 2014, the 101st Airborne established JFC-UA in Liberia.

The U.S. Agency for International Development was the lead U.S. federal agency, and the 101st Airborne deployed to support both them and the government of Liberia. At the peak there were approximately 2,700 JFC-UA personnel in theater, which included personnel from across the U.S. military services.[42] In the course of the mission the JFC-UA supported the training of over 1,500 health care workers and the building and operation of 17 Ebola testing units and the development of logistical systems in-country to move material and medical supplies to the area of greatest need.[43] The 101st Airborne ended the mission and left theater on February 27, 2015, transitioning the tasks to local control, after 5 months of deployment.[44] From notification to deployment on the ground, the operation lasted only 29 days.

---

[36] Andrade, 2010, p. 23.

[37] U.S. Army Office of Public Affairs, 2007.

[38] Andrade, 2010, p. 5.

[39] Andrade, 2010, p. 384.

[40] The White House, Office of the Press Secretary, "Remarks by the President on the Ebola Outbreak," September 16, 2014.

[41] Center for Army Lessons Learned, *101st Airborne Division (Air Assault) Operation United Assistance, Lessons and Best Practices*, Initial Impressions Report 16-05, Washington, D.C., November 2015, p. 5.

[42] Center for Army Lessons Learned, 2015, p. 17.

[43] Center for Army Lessons Learned, 2015, p. iv.

[44] N. Hoskins, "101st Airborne Departs Liberia After Successful Mission," U.S. Department of Defense website, February 27, 2015.

References

Abrams, Robert B.,"FORSCOM Command Training Guidance (CTG)—Fiscal Year 2018," memorandum, Ft. Bragg, N.C., March 24, 2017.

AcqNotes, "JCIDS Process: DOTMLPF-P Analysis." As of May 17, 2018:
http://www.acqnotes.com/acqnote/acquisitions/dotmlpf-analysis

Akgün, Ali E., John Byrne, Halit Keskin, Gary S. Lynn, and Salih Z. Imamoglu, "Knowledge Networks in New Product Development Projects: A Transactive Memory Perspective," *Information and Management*, Vol. 42, No. 8, 2005, pp. 1105–1120.

Andrade, D., *Surging South of Baghdad: The 3D Infantry Division and Task Force Marne in Iraq, 2007–2008*, Washington, D.C.: U.S. Army Center of Military History, 2010.

Army Public Affairs, "Department of the Army Announces 1st Infantry Division Deployment," Army.mil, October 14, 2016.

Austin, John R. "Transactive Memory in Organizational Groups: The Effects of Content, Consensus, Specialization, and Accuracy on Group Performance," *Journal of Applied Psychology*, Vol. 88, No. 5, 2003, p. 866.

Bacevich, J., *The Pentomic Era: The U.S. Army Between Korea and Vietnam*, CreateSpace Independent Publishing Platform, 2012.

Batschelet, Alan, Mike Runey, and Gregory Meyer Jr., "Breaking Tactical Fixation: The Division's Role," *Military Review*, Vol. 89, No. 6, 2009, p. 35.

Bonin, John A., and Telford E. Crisco, Jr., "The Modular Army," *Military Review*, Vol. 84, No. 2, 2004, pp. 21–27.

Brownlee, Les, and Peter J. Schoomaker, *Serving a Nation at War: A Campaign Quality Army with Joint and Expeditionary Capabilities*, Washington, D.C.: Office of the Under Secretary of the Army, 2004.

Bryant, Susan F., *Forging Campaign Quality: Ensuring Adequate Stability Operations Capability within the Modular Army*, Quantico, Va.: Marine Corps University School of Advanced Warfighting, 2006.

Carter, Ashton B., "20% Headquarters Reductions," USNI.org, August 2, 2013. As of July 18, 2017:
https://news.usni.org/2013/08/02/document-carter-memo-on-headquarters-reduction

Center for Army Lessons Learned, *101st Airborne Division (Air Assault) Operation United Assistance, Lessons and Best Practices*, Initial Impressions Report 16-05, Washington, D.C., November 2015.

Dickstein, Corey, "3rd ID Commander Remaining Focused in Afghanistan Ahead of Command Change," *Savannah Morning News*, July 4, 2015. As of September 1, 2017 at:
http://savannahnow.com/news/2015-07-04/3rd-id-commander-remaining-focused-afghanistan-ahead-command-change

———, "Army: 500 from 1st Cavalry Division Deploy to Afghanistan," *Stars and Stripes*, March 22, 2016. As of December 13, 2016:
https://www.stripes.com/news/army-500-from-1st-cavalry-division-deploy-to-afghanistan-1.400621

Fredrick, George L., *METL Task Selection and the Current Operational Environment*, Ft. Leavenworth, Kan.: School of Advanced Military Studies, 2000.

Fukuyama, Francis, and Abram Shulsky, "Military Organization in the Information Age: Lessons from the World of Business," in Zalmay M. Khalilzad and John P. White, eds., *Strategic Appraisal: The Changing Role of Information in Warfare*, Washington, D.C.: RAND Corporation, MR-1016-AF, 1999. As of September 17, 2018: https://www.rand.org/pubs/monograph_reports/MR1016.html

Gates, Robert, Secretary of Defense, "Utilization of Total Force," memorandum to the Secretaries of the Services, Chairman of the Joint Chiefs of Staff and the Undersecretaries of Defense, Washington, D.C., January 19, 2007.

Gino, Francesca, Linda Argote, Ella Miron-Spektor, and Gergana Todorova, "First, Get Your Feet Wet: The Effects of Learning from Direct and Indirect Experience on Team Creativity," *Organizational Behavior and Human Decision Processes*, Vol. 111, No. 2, 2010, pp. 102–115.

Grigsby, Wayne W. Jr., *The Division HQ: Can It Do It All?* Ft. Leavenworth, Kan.: School of Advanced Military Studies, 1996.

Hawkins, Glen R., and James Jay Carafano, *Prelude to Army XXI: U.S. Army Division Design Initiatives and Experiments, 1917–1995*, Washington, D.C.: U.S. Army Center of Military History, 1997.

Headquarters, Department of the Army, FM 71-100, *Division Operations*, Washington, D.C., August 28, 1996.

———, Department of the Army, Field Manual 3-81, *Maneuver Enhancement Brigade*, Washington, D.C., April 21, 2014.

———, Department of the Army, ADP 1-01, Doctrine Primer, Washington. D.C., September 2, 2014.

———, Department of the Army, Army Techniques Publication No. 3-92, *Corps Operations*, Washington D.C., April 7, 2016.

———, Army Pamphlet 220-1, *Defense Readiness Reporting System—Army Procedures*, Washington, D.C., November 16, 2011.

———, FM 7-15, *Army Universal Training List*, Washington, D.C., December 9, 2011.

———, ADRP 7-0, *Training Units and Developing Leaders*, Washington, D.C., August 2012.

———, FM 3-94, *Theater Army, Corps, and Division Operations*, Washington, D.C., April 21, 2014.

———, Army Techniques Publication No. 3-91, *Division Operations* Washington, D.C., October 17, 2014.

———, FM 7-0, *Train to Win in a Complex World,* Washington, D.C., October 5, 2016.

———, Army Techniques Publication No. 6-0.5, *Command Post Organization and Operations*, Washington, D.C., March 1, 2017.

Headquarters, United States Army Forces Command, FORSCOM Regulation 220-2, *Methods for Integrating Regular Army, Army National Guard, and Army Reserve Organizations*, Ft. Bragg, N.C., May 31, 2017.

Hoffman, Frank G., "Complex Irregular Warfare: The Next Revolution in Military Affairs," *Orbis*, Vol. 50, No. 3, 2006, pp. 395–411.

Hoskins, Nathan, "101st Airborne Departs Liberia After Successful Mission," U.S. Department of Defense website, February 27, 2015. As of September 19, 2018:
https://www.defense.gov/News/Article/Article/604187/101st-airborne-departs-liberia-after-successful-mission/

———, "101st Airborne Division Completes Iraq Tour, Transfers Mission to 1st Infantry Division," Army.mil, November 21, 2016. As of December 8, 2016:
https://www.army.mil/article/178618/

Jacobi, K. L., *Division METL—Clinging to an Antiquated Paradigm?* Ft. Leavenworth, Kan.: School of Advanced Military Studies, 2004.

Joint Chiefs of Staff, *CJCS Guide to the Chairman's Readiness System*, CJCS Guide 3401D, Washington, D.C., November 15, 2010.

Joint Chiefs of Staff, JP 1, *Doctrine for the Armed Forces of the United States*, Washington, D.C., March 25, 2013.

Joint Publication 5-0, Department of the Navy and United States Marine Corps, *Joint Operation Planning*, August 2011, p. III- 39.

Jussel, Paul C., *Intimidating the World: The United States Atomic Army, 1956–1960*, dissertation, Columbus, Ohio: The Ohio State University, 2004.

Kedzior, Richard W., *Evolution and Endurance: The U.S. Army Division in the Twentieth Century*, Santa Monica, Calif.: RAND Corporation, MR-1211-A, 2000. As of September 19, 2018:
https://www.rand.org/pubs/monograph_reports/MR1211.html

Kennedy, Christopher, *The US Army Division: The Continuous Evolution to Remain Relevant*, Carlisle, Pa.: U.S. Army War College, 2013.

Kim, Kap, "Division Cases Colors for Upcoming Deployment," Army.mil, October 22, 2015. As of December 13, 2016:
https://www.army.mil/article/157472/

Korpi, Donald, "National Support Element Completes Transfer of Authority from 10th Mountain Infantry Division," Army.mil, September 13, 2016. As of December 13, 2016:
https://www.army.mil/article/175011/

Kretchik, Walter, *US Army Doctrine: From the American Revolution to the War on Terror*, Lawrence: University of Kansas Press, 2012.

Kretchick, W. E., R. F. Baumann, and J. T. Fishel, *Invasion, Intervention, "Intervasion": A Concise History of the U.S. Army in Operation Uphold Democracy*, Ft. Leavenworth, Kan.: U.S. Army Command and General Staff College Press, 1998.

Krulak, Charles C., "The Strategic Corporal: Leadership in the Three Block War," *Marines Magazine*, January 1999. As of September 21, 2017:
http://www.au.af.mil/au/awc/awcgate/usmc/strategic_corporal.htm

Liang, Diane Wei, Richard Moreland, and Linda Argote, "Group Versus Individual Training and Group Performance: The Mediating Role of Transactive Memory," *Personality and Social Psychology Bulletin*, Vol. 21, No. 4, 1995, pp. 384–393.

Littlepage, Glenn, William Robison, and Kelly Reddington, "Effects of Task Experience and Group Experience on Group Performance, Member Ability, and Recognition of Expertise," *Organizational Behavior and Human Decision Processes*, Vol. 69, No. 2, 1997, pp. 133–147.

Lytell, Maria C., Susan G. Straus, Chad C. Serena, Geoffrey Grimm, James L. Doty, Jennie W. Wenger, Andrea A. Golay, Andrew M. Naber, Clifford A. Grammich, and Eric S. Fowler, *Assessing Competencies and Proficiency of Army Intelligence Analysts Across the Career Life Cycle*, Santa Monica, Calif.: RAND Corporation, RR-1851-A, 2017. As of October 24, 2017:
https://www.rand.org/pubs/research_reports/RR1851.html

Macgregor, Douglas A., *Breaking the Phalanx: A New Design for Landpower in the 21st Century*, Ex-library edition, Westport, Conn.: Praeger, 1997.

McHugh, John M., Secretary of the Army, "Army Directive 2012–08 (Army Total Force Policy)," memorandum for principal officials of Headquarters, Department of the Army Commander, Washington, D.C., September 4, 2012.

McNaugher, Thomas L., David E. Johnson, and Jerry M. Sollinger, *Agility by a Different Measure: Creating a More Flexible U.S. Army*, Santa Monica, Calif.: RAND Corporation, IP-195, 2000. As of September 19, 2018:
http://www.rand.org/pubs/issue_papers/IP195.html

Meredith, Lisa S., Carra S. Sims, Benjamin Batorsky, Adeyemi Okunogbe, Brittany L. Bannon, and Craig A. Myatt, *Identifying Promising Approaches to U.S. Army Institutional Change: A Review of the Literature on Organizational Culture and Climate*, Santa Monica, Calif.: RAND Corporation, RR-1588-A, 2016. As of September 19, 2018:
https://www.rand.org/pubs/research_reports/RR1588.html

National Commission on the Future of the Army, *Report to the President and the Congress of the United States*, Arlington, Va.: U.S. Department of Defense, 2016.

Neumann, B. F., L. Mundey, and J. Mikolashek, *The U.S. Army in Afghanistan Operation Enduring Freedom, March 2002–April 2005*, Washington, D.C.: Center for Military History, United States Army, 2013.

Pernin, Christopher G., Katharina Ley Best, Matthew E. Boyer, Jeremy M. Eckhause, John Gordon IV, Dan Madden, Katherine Pfrommer, Anthony D. Rosello, Michael Schwille, Michael Shurkin, and Jonathan P. Wong, *Enabling the Global Response Force Access Strategies for the 82nd Airborne Division*, Santa Monica, Calif.: RAND Corporation, RR-1161-A, 2017. As of July 31, 2017:
https://www.rand.org/pubs/research_reports/RR1161.html

Pint, Ellen M., Matthew W. Lewis, Thomas F. Lippiatt, Philip Hall-Partyka, Jonathan P. Wong, and Tony Puharic, *Active Component Responsibility in Reserve Component Pre- and Postmobilization Training*, Santa, Monica, Calif.: RAND Corporation, RR-738-A, 2015. As of September 19, 2018:
https://www.rand.org/pubs/research_reports/RR738.html

Pint, Ellen M., Christopher M. Schnaubelt, Stephen Dalzell, Jaime Hastings, Penelope Speed, and Michael G. Shanley, *Review of Army Total Force Policy Implementation*, Santa Monica, Calif.: RAND Corporation, RR-1958-A, 2017. As of September 19, 2018:
https://www.rand.org/pubs/research_reports/RR1958.html

Ren, Yuqing, Kathleen M. Carley, and Linda Argote, "The Contingent Effects of Transactive Memory: When Is It More Beneficial to Know What Others Know?" *Management Science*, Vol. 52, No. 5, 2006, pp. 671–682.

Rulke, Diane Liang, and Devaki Rau, "Investigating the Encoding Process of Transactive Memory Development in Group Training," *Group and Organization Management*, Vol. 25, No. 4, 2000, pp. 373–396.

Schein, Edgar H., "Organizational Culture," *American Psychologist*, Vol. 45, No. 2, 1990, pp. 109–119.

Schnaubelt, Christopher M., Raphael S. Cohen, Molly Dunigan, Gian Gentile, Jaime L. Hastings, Joshua Klimas, Jeffrey P. Marquis, Agnes G. Schaefer, Bonnie Triezenberg, and Michelle Darrah Ziegler, *Sustaining the Army's Reserve Components as an Operational Force*, Santa Monica, Calif.: RAND Corporation, RR-1495-A, 2016. As of August 4, 2017:
https://www.rand.org/pubs/research_reports/RR1495.html

Secretary of Defense, Memorandum for Secretaries of the Military Departments, Chairman of the Joint Chiefs of Staff, and Under Secretaries of Defense, "Utilization of the Total Force," January 19, 2007. As of September 19, 2018:
http://www.armyg1.army.mil/MilitaryPersonnel/Hyperlinks/Adobe%20Files/OSD%20Memo%20dtd%2020070119%20-%20Utilization%20of%20the%20Force.pdf

Swenddal, Jerem G., and Stacy L. Moore, "From Riley to Baku: How an Opportunistic Unit Broke the Crucible," *Military Review*, January–February 2017. As of August 2, 2017:
http://www.armyupress.army.mil/Journals/Military-Review/English-Edition-Archives/January-February-2017/ART-012/

Tan, Michelle, "3rd ID Commander Readies His Troops for Afghanistan," ArmyTimes.com, November 9, 2014. As of January 25, 2016:
https://www.armytimes.com/story/military/2014/11/09/3rd-id-commander-readies-his-troops-for-afghanistan/18767595/

———, "101st Airborne to Deploy to Iraq, Kuwait," ArmyTimes.com, November 6, 2015. As of September 1, 2017:
http://www.armytimes.com/news/your-army/2015/11/06/101st-airborne-to-deploy-to-iraq-kuwait/

———, "'Objective T': The Army's New Mission to Track Training," *Army Times*, October 11, 2016. As of February 1, 2019:
https://www.armytimes.com/news/your-army/2016/10/11/objective-t-the-army-s-new-mission-to-track-training/

Truong, Chi, "48th CBRN Brigade Uncases Colors in Liberia," *Fort Hood Sentinel*, April 2, 2015.

U.S. Army, *Military Occupation and Classification Structure*, Department of the Army Pamphlet, 611-21, Washington, D.C.: U.S. Army, August 10, 2008.

U.S. Army Acquisition Support Center, "Distributed Common Ground System–Army," Army.mil, n.d. As of September 19, 2018:
http://asc.army.mil/web/portfolio-item/iews-dcgs-a/

U. S. Army Combined Arms Center, "Corps and Division Redesign (FARG II) Force Design Update Brief," December 2015.

———, *Division and Corps Reduction (FARG II) Organizational Design Paper*, April 15, 2016.

U.S. Army General Staff, *Field Service Regulations United States Army 1905: With Amendments to 1908*, Washington, D.C.: Government Printing Office, 1908, sec. 1.

U.S. Army Office of Public Affairs, "Headquarters 3rd Infantry Division—'Rock of the Marne'—Goes to Iraq in March," Army.mil, February 16, 2007. As of August 30, 2017:
https://www.army.mil/article/1876/Headquarters_3rd_Infantry_Division____quot_Rock_of_the_Marne_quot____Goes_to_Iraq_in_March

U.S. Army War College, *How the Army Runs 2015–2016: A Senior Leader Reference Handbook*, Carlisle, Pa.: U.S. Army War College, 2015.

U.S. Department of Defense, *United States Forces, Somalia After Action Report and Historical Overview: The United States Army in Somalia, 1992–1994*, Washington, D.C.: U.S. Army Center of Military History, 2003.

———, "Managing the Reserve Components as an Operational Force," directive no. 1200.17, October 29, 2008.

Vergun, David, "101st HQ Deploying to Liberia in Response to Ebola Epidemic," Army.mil, September 30, 2014. As of January 30, 2019:
https://www.army.mil/article/134936/101st_hq_deploying_to_liberia_in_response_to_ebola_epidemic

Vidal, James. "Engineer Battalion Deploys Soldiers in Fight Against Ebola," Army.mil, October 17, 2014. As of March 22, 2016:
http://www.army.mil/article/13616

Weigley, Russell Frank, *History of the United States Army*, New York: Macmillan, 1967.

The White House, Office of the Press Secretary, "Remarks by the President on the Ebola Outbreak," September 16, 2014. As of September 19, 2018:
https://obamawhitehouse.archives.gov/the-press-office/2014/09/16/remarks-president-ebola-outbreak

Wilson, John, B., *Maneuver and Firepower: The Evolution of Divisions and Separate Brigades*, Washington, D.C.: U.S. Army Center of Military History, 1998.

Wilson, John B., "Mobility Versus Firepower: The Post-World War I Infantry Division," Parameters, Vol. 13, No. 3, 1983, p. 47.

Wilson, Peter A., John Gordon IV, and David E. Johnson, "An Alternative Future Force: Building a Better Army," *Parameters*, Vol. 33, No. 4, 2003.

Wright, Robert K., *The Continental Army*, Washington, D.C.: U.S. Army Center of Military History, 1989.